John Tyndall

Lectures on Light

Delivered in the United States in 1872-'73

John Tyndall

Lectures on Light
Delivered in the United States in 1872-'73

ISBN/EAN: 9783337252052

Printed in Europe, USA, Canada, Australia, Japan

Cover: Foto ©berggeist007 / pixelio.de

More available books at **www.hansebooks.com**

LECTURES ON LIGHT.

DELIVERED IN THE UNITED STATES

IN 1872-'73.

BY

JOHN TYNDALL, LL. D., F. R. S.,

PROFESSOR OF NATURAL PHILOSOPHY IN THE ROYAL INSTITUTION.

WITH AN APPENDIX.

NEW YORK:
D. APPLETON AND COMPANY,
549 AND 551 BROADWAY.
1873.

ENTERED, according to Act of Congress, in the year 1873,
By D. APPLETON & COMPANY,
In the Office of the Librarian of Congress, at Washington.

PREFACE.

My eminent friend Prof. Joseph Henry, of Washington, did me the honor of taking these lectures under his personal direction, and of arranging the times and places at which they were to be delivered.

Deeming that my home-duties could not, with propriety, be suspended for a longer period, I did not, at the outset, expect to be able to prolong my visit to the United States beyond the end of 1872.

Thus limited as to time, Prof. Henry began in the North, and, proceeding southwards, arranged for the successive delivery of the lectures in Boston, New York, Philadelphia, Baltimore, and Washington.

By this arrangement, which circumstances at the time rendered unavoidable, the lectures in New York were rendered coincident with the period of the presidential election. This was deemed unsatisfactory, and when the fact was represented to me I at once offered to extend the time of my visit so as to make the lectures in New York succeed those in Washington. The proposition was cordially accepted by my friends.

To me personally this modified arrangement has proved in the highest degree satisfactory. It gave me a much-needed holiday at Niagara Falls; it, moreover, rendered the successive stages of my work a kind of *growth*, which reached its most impressive development in New York and Brooklyn.

In every city that I have visited, my reception has been that of a friend; and, now that my visit has become virtually a thing of the past, I can look back upon it with unqualified pleasure. It is a memory without a stain—an experience of deep and genuine kindness on the part of the American people, never, on my part, to be forgotten.

This relates to what may be called the *positive* side of my visit—to the circumstances attending the work actually done. My only drawback relates to work *undone;* for I carry home with me the consciousness of having been unable to respond to the invitations of the great cities of the West; thus, I fear, causing, in many cases, disappointment. Would that this could have been averted! But the character of the lectures, and the weight of instrumental appliances which they involved, entailed loss of time and heavy labor. The need of rest alone would be a sufficient admonition to me to pause here; but, besides this, each successive mail from London brings me intelligence of work suspended and duties postponed through my absence. These are the considerations which prevent me from responding, with a warmth commensurate with their own, to the wishes of my friends in the West.

On quitting England, I had no intention of publishing these lectures, and, except a fragment or two, not a line of them was written when I reached this city. They have been begun, continued, and ended, in New York, and bear only too evident marks of the rapidity of their production. I thought it, however, due, both to those who heard them with such marked attention, and to those who wished to hear them, but were unable to do so, to leave them behind me in an authentic form. The execution of this work has cut me off from many social pleasures; it has also prevented me from making myself acquainted with institutions in the working of which I feel a deep interest. But human power is finite, and mine has been expended in the way which I deemed most agreeable, not to my more intimate friends, but to the people of the United States.

In the opening lecture are mentioned the names of gentlemen to whom I am under lasting obligations for their friendly and often laborious aid. The list might readily be extended, for in every city I have visited willing helpers were at hand. I must not, however, omit the name of Mr. Rhees, Professor Henry's private secretary, who not only in Washington, but in Boston, gave me most important assistance. To the trustees of the Cooper Institute my acknowledgments are due; also to the directors of the Mercantile Library at Brooklyn. I would add to these a brief but grateful reference to my high-minded friend and kinsman, General Hector Tyndale, for his long-continued care of me, and for the thoughtful tenderness by which he and his family softened, both to me and to the parents of the youth, the pain occasioned by the death of my junior assistant in Philadelphia.

Finally, I have to mention with warm commendation the integrity, ability, and devotion, with which, from first to last, I have been aided by my principal assistant, Mr. John Cottrell.

New York, *February*, 1873.

CONTENTS.

LECTURE I.

INTRODUCTORY: Uses of Experiment: Early Scientific Notions: Sciences of Observation: Knowledge of the Ancients regarding Light: Nature judged from Theory defective: Defects of the Eye: Our Instruments: Rectilineal Propagation of Light: Law of Incidence and Reflection: Sterility of the Middle Ages: Refraction: Discovery of Snell: Descartes and the Rainbow: Newton's Experiments on the Composition of Solar Light: His Mistake as regards Achromatism: Synthesis of White Light: Yellow and Blue Lights proved to produce White by their Mixture: Colors of Natural Bodies: Absorption: Mixture of Pigments contrasted with Mixture of Lights,   p. 9

LECTURE II.

Origin of Physical Theories: Scope of the Imagination: Newton and the Emission Theory: Verification of Physical Theories: The Luminiferous Ether: Wave-Theory of Light: Thomas Young: Fresnel and Arago: Conceptions of Wave-Motion: Interference of Waves: Constitution of Sound-Waves: Analogies of Sound and Light: Illustrations of Wave-Motion: Interference of Sound-Waves: Optical Illustrations: Pitch and Color: Lengths of the Waves of Light and Rates of Vibration of the Ether-Particles: Interference of Light: Phenomena which first suggested the Undulatory Theory: Hooke and the Colors of Thin Plates: The Soap-Bubble: Newton's Rings: Theory of "Fits:" Its Explanation of the Rings: Overthrow of the Theory: Colors of Mother-of-Pearl, ... p. 33

LECTURE III.

Relation of Theories to Experience: Origin of the Notion of the Attraction of Gravitation: Notion of Polarity, how generated: Atomic Polarity: Structural Arrangements due to Polarity: Architecture of Crystals considered as an Introduction to their Action upon Light: Notion of Atomic Polarity applied to Crystalline **Structure**: Experimental Illustrations: **Crystallization of Water**: **Expansion by Heat and by Cold**: Deportment of Water considered and explained: Molecular Action Illustrated by a Model: Force of Solidification: Bearings of Crystallization on Optical Phenomena: Refraction: Double Refraction: Polarization: Action of Tourmaline: Character of the Beams emergent from Iceland Spar: Polarization by ordinary Refraction and Reflection: Depolarization, ... p. 69

LECTURE IV.

Chromatic Phenomena produced by Crystals on Polarized Light: The Nicol Prism: Polarizer and Analyzer: Action of thick and thin Plates of Selenite: Colors dependent on Thickness: Resolution of Polarized Beam into two others by the Selenite: One of them more retarded than the other: Recompounding of the two Systems of Waves by the Analyzer: Interference thus rendered possible: Consequent Production of Colors: Action of Bodies Mechanically strained or pressed: Action of Sonorous Vibrations: Action of Glass strained or pressed by Heat: Circular Polarization: Chromatic Phenomena produced by Quartz: The Magnetization of Light: Rings surrounding the Axes of Crystals: Biaxal and Uniaxal Crystals: Grasp of the Undulatory Theory, ... p. 101

LECTURE V.

Range of Vision incommensurate with Range of Radiation: The Ultra-Violet Rays: Fluorescence: Rendering Invisible Rays visible: Vision not the only Sense appealed to by the Solar and Electric Beam: Heat of Beam: Combustion by Total Beam at the Foci of Mirrors and Lenses: Combustion through Ice-Lens: Ignition of Diamond: Search for the Rays here effective: Sir William Herschel's Discovery of Dark Solar Rays: Invisible Rays the Basis of the Visible: Detachment by a Ray-Filter of the Invisible Rays from the Visible: Combustion at Dark Foci: Conversion of Heat-Rays into Light-Rays: Calorescence: Part played in Nature by Dark Rays: Identity of Light and Radiant Heat: Invisible Images: Reflection, Refraction, Plane Polarization, Depolarization, Circular Polarization, Double Refraction, and Magnetization of Radiant Heat, ... p. 127

LECTURE VI.

Principles of Spectrum Analysis: Solar Chemistry: Summary and Conclusion,
p. 151

APPENDIX: Address of Prof. Tyndall at the Farewell Banquet, . p. 184

LECTURE I.

INTRODUCTORY: Uses of Experiment: Early Scientific Notions: Sciences of Observation: Knowledge of the Ancients regarding Light: Nature judged from Theory defective: Defects of the Eye: Our Instruments: Rectilineal Propagation of Light: Law of Incidence and Reflection: Sterility of the Middle Ages: Refraction: Discovery of Snell: Descartes and the Rainbow: Newton's Experiments on the Composition of Solar Light: His Mistake as regards Achromatism: Synthesis of White Light: Yellow and Blue Lights proved to produce White by their Mixture: Colors of Natural Bodies: Absorption: Mixture of Pigments contrasted with Mixture of Lights.

SOME twelve years ago I published, in England, a little book entitled the "Glaciers of the Alps," and, a couple of years subsequently, a second volume, entitled "Heat as a Mode of Motion." These volumes were followed by others, written with equal plainness, and with a similar aim, that aim being to develop and deepen sympathy between science and the world outside of science. I agreed with thoughtful men [1] who deemed it good for neither world to be isolated from the other, or unsympathetic towards the other, and, to lessen this isolation, at least in one department of science, I swerved aside from those original researches which had previously been the pursuit and pleasure of my life.

[1] Among whom may be mentioned, specially, the late Sir Edmund Head, Bart.

These books were, for the most part, republished by the Messrs. Appleton, under the auspices of a man who is untiring in his efforts to diffuse sound scientific knowledge among the people of this country; whose energy, ability, and single-mindedness, in the prosecution of an arduous task, have won for him the sympathy and support of many of us in "the old country." I allude to Professor Youmans, of this city. Quite as rapidly as in England, the aim of these works was understood and appreciated in the United States, and they brought me from this side of the Atlantic innumerable evidences of good-will. Year after year invitations reached me [1] to visit America, and last year I was honored with a request so cordial, and signed by five-and-twenty names so distinguished in science, in literature, and in administrative position, that I at once resolved to respond to it by braving not only the disquieting oscillations of the Atlantic, but the far more disquieting ordeal of appearing in person before the people of the United States.

This request, conveyed to me by my accomplished friend Professor Lesley, of Philadelphia, and preceded by a letter of the same purport from your scientific Nestor, Professor Joseph Henry, of Washington, desired that I would lecture in some of the principal cities of the Union. This I agreed to do, though much in the dark

[1] One of the earliest came from Mr. John Amory Lowell, of Boston.

as to what form such lectures ought to take. In answer to my inquiries, however, I was given to understand (by Professor Youmans principally) that a course of experimental lectures would materially promote scientific education in this country, and I at once resolved to meet this desire as far as my time allowed.

Experiments have two uses—a use in discovery, and a use in tuition. They are the investigator's language addressed to Nature, to which she sends intelligible replies. These replies, however, are, for the most part, at first too feeble for the public ear; for the investigator cares little for the loudness of Nature's voice if he can only unravel its meaning. But after the discoverer comes the teacher, whose function it is so to exalt and modify the results of the discoverer as to render them fit for public presentation. This secondary function I shall endeavor, in the present instance, to fulfil.

I propose to take a single department of natural philosophy, and illustrate, by means of it, the growth of scientific knowledge under the guidance of experiment. I wish, in this first lecture, to make you acquainted with certain elementary phenomena; then to point out to you how those theoretic principles by which phenomena are explained, take root, and flourish in the human mind, and afterwards to apply these principles to the whole body of knowledge covered by the lectures. The science of optics lends itself to this mode of treatment, and on it, therefore, I propose to draw for the materials of the present course. It will be best to begin with the few simple facts regarding light which were known to the ancients, and to pass from them in historic gradation to the more abstruse discoveries of modern times.

All our notions of Nature, however exalted or however grotesque, have some foundation in experience. The notion of personal volition in Nature had this basis. In the fury and the serenity of natural phenomena the savage saw the transcript of his own varying moods, and he accordingly ascribed these phenomena to beings of like passions with himself, but vastly transcending him in power. Thus the notion of *causality*—the assumption that natural things did not come of themselves, but had unseen antecedents—lay at the root of even the savage's interpretation of Nature. Out of this bias of the human mind to seek for the antecedents of phenomena all science has sprung.

The development of man, indeed, is ultimately due to his interaction with Nature. Natural phenomena arrest his attention and excite his questionings, the intellectual activity thus provoked reacting on the intellect itself, and adding to its strength. The quantity of power added by any single effort of the intellect may be indefinitely small; but the integration of innumerable increments of this kind has raised intellectual power from its rudiments to the magnitude it possesses to-day. In fact, the indefinite smallness of the single increment is made good by the indefinite number of such increments, summed up in what may be regarded as practically infinite time.

We will not now go back to man's first intellectual gropings; much less shall we enter upon the thorny discussion as to how the groping man arose. We will take him at a certain stage of his development, when, by evolution or sudden endowment, he became possessed of the apparatus of thought and the power of using it. For a time—and that historically a long one—he was limited to mere observation, accepting what Nature offered, and confining intellectual action to it. The apparent motions of sun and stars first drew towards them the questionings of the intellect, and accordingly astronomy was the first science developed. Slowly, and with difficulty, the notion of natural forces took root in the mind, the seedling of this notion being the actual observation of electric and magnetic attractions. Slowly, and with difficulty, the science of mechanics had to grow out of this notion; and slowly at last came the full application of mechanical principles to the motions of the heavenly bodies. We trace the progress of astronomy through Hipparchus and Ptolemy; and, after a long halt, through Copernicus, Galileo, Tycho Brahe, and Kepler; while from the high table-land of thought raised by these men Newton shoots upward like a peak, overlooking all others from his dominant elevation.

But other objects than the motions of the stars attracted the attention of the ancient world. Light was a familiar phenomenon, and from the earliest times we find men's minds busy with the attempt to render some account of it. But without experiment, which belongs to a later stage of scientific development, little progress could be made in this subject. The ancients, accordingly, were far less successful in dealing with light than in dealing with solar and stellar motions. Still they did make some progress. They satisfied themselves that light moved in straight lines; they knew also that these lines or *rays* of light were reflected from polished surfaces, and that the angle of incidence was equal to the angle of reflection. These two results of ancient scientific curiosity constitute the starting-point of our present course of lectures.

But in the first place it may be useful to say a few words regarding the source of light to be employed in our experiments. The rusting of iron is, to all intents and purposes, the slow burning of iron. It develops heat, and, if the heat be preserved, a high temperature may be thus attained. The destruction of the first Atlantic cable was probably due to heat developed in this way. Other metals are still more combustible than iron. You may light strips of zinc in a candle-flame, and cause them to burn almost like strips of paper. But, besides combustion in the air, we may also have combustion in a liquid. Water, for example, contains a store of oxygen, which may unite with and consume a metal immersed in it. It is from this kind of combustion that we are to derive the heat and light employed in the present course.

Their generation merits a moment's attention. Before you is an instrument—a small voltaic battery—in which zinc is immersed in a suitable liquid. Matters are so arranged that an attraction is set up between the metal and the oxygen, actual union, however, being in the first instance avoided. Uniting the two ends of the battery by a thick wire, the attraction is satisfied, the oxygen unites with the metal, the zinc is consumed, and heat, as usual, is the result of the combustion. A power, which, for want of a better name, we call an electric current, passes at the same time through the wire.

Cutting the thick wire in two, I unite the severed ends by a thin one. It glows with a white heat. Whence comes that heat? The question is well worthy of an answer. Suppose in the first instance, when the thick wire was employed, that we had permitted the action to continue until 100 grains of zinc were consumed, the amount of heat generated in the battery would be capable of accurate numerical expression. Let the action now continue, with this thin wire glowing, until 100 grains of zinc are consumed. Will the amount of heat generated in the battery be the same as before? No, it will be less by the precise amount generated in the thin wire outside the battery. In fact, by adding the internal heat to the external, we obtain for the combustion of 100 grains of zinc a total which never varies. By this arrangement, then, we are able to burn our zinc at one place, and to exhibit the heat and light of its combustion at a distant place. In New York, for example, we have our grate and fuel; but the heat and light of our fire may be made to appear at San Francisco.

I now remove the thin wire and attach to the severed ends of the thick one two thin rods of coke. On bringing the rods together we obtain a small star of light. Now, the light to be employed in our lectures is a simple exaggeration of this star. Instead of being produced by ten cells, it is produced by fifty. Placed in a suitable camera, provided with a suitable lens, this light will give us all the beams necessary for our experiments.

And here, in passing, let me refer to the common delusion that the works of Nature, the human eye included, are theoretically perfect. The degree of perfection of any organ is determined by what it has to do. Looking at the dazzling light from our large battery, you see a globe of light, but entirely fail to see the shape of the coke-points whence the light issues.

The cause may be thus made clear: On the screen before you is projected an image of the carbon-points, the *whole* of the lens in front of the camera being employed to form the image. It is not sharp, but surrounded by a halo which nearly obliterates it. This arises from an imperfection of the lens, called its *spherical aberration*, due to the fact that the circumferential and central rays have not the same focus. The human eye labors under a similar defect, and, when you looked at the naked light from fifty cells, the blur of light upon the retina was sufficient to destroy the definition of the retinal image of the carbons. A long list of indictments might indeed be brought against the eye—its opacity, its want of symmetry, its lack of achromatism, its absolute blindness, in part. All these taken together caused an eminent German philosopher to say that, if any optician sent him an instrument so full of defects, he would send it back to him with the severest censure. But the eye is not to be judged from the stand-point of theory. As a practical instrument, and taking the adjustments by which its defects are neutralized into account, it must ever remain a marvel to the reflecting mind.

The ancients, as I have said, were aware of the rectilineal propagation of light. They knew that an opaque body, placed between the eye and a point of light, intercepted the light of the point. Possibly the terms " ray " and " beam " may have been suggested by those straight spokes of light which, in certain states of the atmosphere, dart from the sun at his rising and his setting. The rectilineal propagation of light may be illustrated at home in this way: Make a small hole in a closed window-shutter, before which stands a house or a tree, and place within the darkened room a white screen at some distance from the orifice. Every straight ray proceeding from the house or tree stamps its color upon the screen, and the sum of all the rays forms an image of the object. But, as the rays cross each other at the orifice, the image is inverted. Here we may illustrate the subject thus: In front of our camera is a large opening, closed at present by a sheet of tin-foil. Pricking by means of a common sewing-needle a small aperture in the tin-foil, an inverted image of the carbon-points starts forth upon the screen. A dozen apertures will give a dozen images, a hundred a hundred, a thousand a thousand. But, as the apertures come closer to each other, that is to say, as the tin-foil between the apertures vanishes, the images overlap more and more. Removing the tin-foil altogether, the screen becomes uniformly illuminated. Hence the light upon the screen may be regarded as the overlapping of innumerable images of the carbon-points. In like manner the light upon every white wall on a cloudless day may be regarded as produced by the superposition of innumerable images of the sun.

The law that the angle of incidence is equal to the

angle of reflection is illustrated in this simple way: A straight lath is placed as an index perpendicular to a small looking-glass capable of rotation. A beam of light is received upon the glass and reflected back along the line of its incidence. Though the incident and the reflected beams pass in opposite directions, they do not jostle or displace each other. The index being turned, the mirror turns along with it, and at each side of the index the incident and the reflected beams are seen tracking themselves through the dust of the room. The mere inspection of the two angles enclosed between the index and the two beams suffices to show their equality. The same simple apparatus enables us to illustrate a law of great practical importance, namely, that, when a mirror rotates, the angular velocity of a beam reflected from it is twice that of the reflecting mirror. One experiment will make this plain to you. The mirror is now vertical, and both the incident and reflected beams are horizontal. Turning the mirror through an angle of 45° the reflected beam is vertical; that is to say, it has moved 90°, or through twice the angle of the mirror.

One of the problems of science, on which scientific progress mainly depends, is to help the senses of man by carrying them into regions which could never be attained without such help. Thus we arm the eye with the telescope when we want to sound the depths of space, and with the microscope when we want to explore motion and structure in their infinitesimal dimensions. Now, this law of angular reflection, coupled with the fact that a beam of light possesses no weight, gives us the means of magnifying small motions to an extraordinary degree. Thus, by attaching mirrors to his suspended magnets, and by watching the images of scales reflected from the mirrors, the celebrated Gauss was able to detect the slightest thrill of variation on the part of the earth's magnetic force. The minute elongation of a bar of metal by the mere warmth of the hand may be so magnified by this method as to cause the index-beam to move from the ceiling to the floor of this room. The elongation of a bar of iron when it is magnetized may be thus demonstrated. By a similar arrangement the feeble attractions and repulsions of the diamagnetic force have been made manifest; while in Sir William Thompson's reflecting galvanometer the principle receives one of its latest applications.

For more than 1,000 years no step was taken in optics beyond this law of reflection. The men of the Middle Ages, in fact, endeavored on the one hand to develop the laws of the universe out of their own consciousness, while many of them were so occupied with the concerns of a future world that they looked with a lofty scorn on all things pertaining to this one. Notwithstanding its demonstrated failure during 1,500 years of trial, there are still men among us who think the riddle of the universe is to be solved by this appeal to consciousness. And, like most people who support a delusion, they maintain theirs warmly, and show scant respect for those who dissent from their views.[1] As regards the refraction of light, the course of real inquiry was resumed in 1100 by an Arabian philosopher named Alhazen. Then it was taken up in succession by Roger Bacon, Vitellio, and Kepler. One of the most important occupations of science is the determination, by precise measurements, of the quantitative relations of phenomena. The value of such measurements depends upon the skill and conscientiousness of the man who makes them. Vitellio appears to have been both skilful and conscientious, while Kepler's habit was to rummage through the observations of his predecessors, look at them in all lights, and thus distil from them the principles which united them. He had done this with the astronomical measurements of Tycho Brahe, and had extracted from them the celebrated "laws of Kepler." He did it also with the measurements of Vitellio. But in the case of refraction he was not successful. The principle, though a simple

---

[1] Schelling thus expresses his contempt for experimental knowledge: "Newton's Optics is the greatest illustration of a whole structure of fallacies, which in all its parts is founded on observation and experiment." There are some small imitators of Schelling still in Germany.

one, escaped him. It was first discovered by Willebrod Snell, about the year 1621.

Less with the view of dwelling upon the phenomenon itself than of introducing it to you in a form which will render intelligible the play of theoretic thought in Newton's mind, I will show you the fact of refraction. The dust of the air and the turbidity of a liquid may here be turned to account. A shallow circular vessel with a glass face, half filled with water, rendered barely turbid by the precipitation of a little mastic, is placed upon its edge with its glass face vertical. Through a slit in the hoop surrounding the vessel a beam of light is admitted. It impinges upon the water, enters it, and tracks itself through the liquid in a sharp, bright band. Meanwhile the beam passes unseen through the air above the water, for the air is not competent to scatter the light. A puff of tobacco-smoke into this space at once reveals the track of the incident-beam. If the incidence be vertical, the beam is unrefracted. If oblique, its refraction at the common surface of air and water is rendered clearly visible. It is also seen that *reflection* accompanies refraction, the beam dividing itself at the point of incidence into a refracted and a reflected portion.

The law by which Snell connected together all the measurements executed up to his time, is this: Let A B C D represent the outline of our circular vessel (Fig. 1), A C being the water-line. When the beam

SNELL'S LAW OF REFRACTION. 23

is incident along B E, which is perpendicular to A C, there is no refraction. When it is incident along m E, there is refraction: it is bent at E and strikes the circle at n. When it is incident along m' E, there is also refraction at E, the beam striking the point n'. From the ends of the incident beams, let the perpendiculars $m\ o$, $m'\ o'$ be drawn upon B D, and from the ends of the refracted beams let the perpendiculars

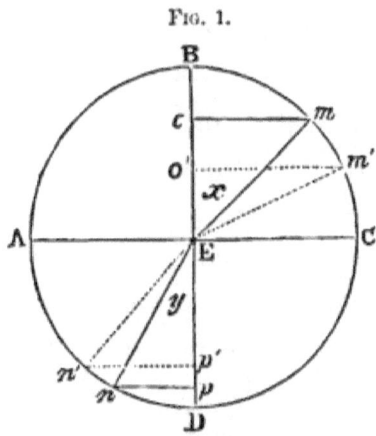

Fig. 1.

$p\ n$, $p'\ n'$ be also drawn. Measure the lengths of $o\ m$ and of $p\ n$, and divide the one by the other. You obtain a certain quotient. In like manner divide $m'\ o'$ by the corresponding perpendicular $p'\ n'$; you obtain in each case *the same quotient.* Snell, in fact, found this quotient to be *a constant quantity* for each particular substance, though it varied in amount from substance to substance. He called the quotient the *index of refraction.*

This law is one of the corner-stones of optical science, and its applications to-day are million-fold. Immediately after its discovery Descartes applied it to the explanation of the rainbow. The bow is seen when the back is turned to the sun. Draw a straight line through the spectator's eye and the sun, the bow is always seen at the same angular distance from this line. This was the great difficulty. Why should the bow be always, and at all its parts, forty one degrees distant from this line? Taking a pen and calculating the track of every ray through a rain-drop, Descartes found that, at one particular angle, the rays emerged from the drop almost parallel to each other; being thus enabled to preserve their intensity through long atmospheric distances; at all other angles the rays quitted the drop *divergent*, and through this divergence became so enfeebled as to be practically lost to the eye. The particular angle here referred to was the foregoing angle of forty-one degrees, which observation had proved to be invariably that of the rainbow.

But in the rainbow a new phenomenon was introduced—the phenomenon of color. And here we arrive at one of those points in the history of science, when men's labors so intermingle, that it is difficult to assign to each worker his precise meed of honor. Descartes was at the threshold of the discovery of the composition of solar light. But he failed to attain perfect clearness, and it is certain that he did not enunciate the true law. This was reserved for Newton, who went to work in this way: Through the closed window-shutter of a room he pierced an orifice, and allowed a thin sunbeam to pass through it. The beam stamped a round image of the sun on the opposite white wall of the room. In the path of this beam Newton placed a prism, expecting to see the beam refracted, but also expecting to see the image of the sun, after refraction, round; to his astonishment, it was drawn out to an image whose length was five times its breadth; and this image was divided into bands of different colors. Newton saw immediately that solar light was *composite*, not simple. His image revealed to him the fact that some constituents of the solar light were more deflected by the prism than others, and he concluded, therefore, that white solar light was a mixture of lights of different colors and of different degrees of refrangibility

Let us reproduce this celebrated experiment. On the screen is now stamped a luminous disk, which may stand for Newton's image of the sun. Causing the beam which produces the disk to pass through a prism, we obtain Newton's elongated colored image, which he called *a spectrum*. Newton divided the spectrum into seven parts, red, orange, yellow, green, blue, indigo, violet; which are commonly called the seven primary or prismatic colors. This drawing out of the white light into its constituent colors is called *dispersion*.

This was the first analysis of solar light by Newton; but the scientific mind is fond of verification, and never neglects it where it is possible. It is this stern conscientiousness in testing its conclusions that gives adamantine strength to science, and renders all assaults on it unavailing. Newton completed his proof by synthesis in this way: The spectrum now before you is produced by a glass prism. Causing the decomposed beam to pass through a second similar prism, but so placed that the colors are refracted back and reblended, the perfectly white image of the slit is restored. Here, then, refraction and dispersion are simultaneously abolished. Are they always so? Can we have the one without the other? It was Newton's conclusion that we could not. Here he erred, and his error, which he maintained to the end of his life, retarded the progress of optical discovery. Dolland subsequently proved that, by combining two different kinds of glass, the colors could be extinguished, still leaving a residue of refraction, and he employed this residue in the construction of achromatic lenses—lenses which yield no color—which Newton thought an impossibility. By setting a water-prism—water contained in a wedge-shaped vessel with glass sides—in opposition to a prism of glass, this point can be illustrated before you. We have first the position of the unrefracted beam marked upon the screen; then we produce the water-spectrum; finally, by introducing a flint-glass prism, we refract the beam back, until the color disappears. The image of the slit is now *white ;* but you see that, though the dispersion is abolished, the refraction is not.

This is the place to illustrate another point bearing upon the instrumental means employed in these lectures. Note the position of the water-spectrum upon the screen. Altering, in no particular, the wedge-shaped vessel, but simply substituting for the water the transparent bisulphide of carbon, you notice how much higher the beam is thrown, and how much richer is the display of color. This will explain to you the use of this substance in our subsequent experiments.

The synthesis of white light may be effected in three ways, which are now worthy of special attention: Here, in the first instance, we have a rich spectrum produced by a prism of bisulphide of carbon. One face of the prism is protected by a diaphragm with a longitudinal slit, through which the beam passes into the prism. It emerges decomposed at the other side. I permit the colors to pass through a cylindrical lens, which so squeezes them together as to produce upon the screen a sharply-defined rectangular image of the longitudinal slit. In that image the colors are re-blended, and you see it perfectly white. Between the prism and the cylindrical lens may be seen the colors tracking themselves through the dust of the room. Cutting off the more refrangible fringe by a card, the rectangle is seen red; cutting off the less refrangible fringe, the rectangle is seen blue. By means of a thin glass prism, I deflect one portion of the colors, and leave the residual portion. On the screen are now two colored rectangles produced in this way. These are *complementary* colors—colors which, by their union, produce white. Note that, by judicious management, one of these colors is rendered *yellow*, and the other *blue*. I withdraw the thin prism; yellow falls upon blue, and we have *white* as the result of their union. On our way, we thus abolish the fallacy, first exposed by Helmholtz, that the mixture of blue and yellow lights produces green.

Again, restoring the circular aperture, we obtain once more a spectrum like that of Newton. By means of a lens, we gather up these colors, and build them together, not to an image of the aperture, but to an image of the carbon-points themselves. Finally, in virtue of the persistence of impressions upon the retina, by means of a rotating disk, on which are spread in sectors the colors of the spectrum, we blend together the prismatic colors *in the eye itself*, and thus produce the impression of whiteness.

Having unravelled the interwoven constituents of white light, we have next to inquire, What part the constitution so revealed enables this agent to play in Nature? To it we owe all the phenomena of color; and yet not to it alone, for there must be a certain relationship between the ultimate particles of natural bodies and light to enable them to extract from it the luxuries of color. But the function of natural bodies is here *selective*, not *creative*. There is no color generated by any natural body whatever. Natural bodies have showered upon them, in the white light of the sun, the sum total of all possible colors, and their action is limited to the sifting of that total, the appropriating from it of the colors which really belong to them, and the rejecting of those which do not. It will fix this subject in your minds if I say that it is the portion of light which they reject, and not that which belongs to them, that gives bodies their colors.

Let us begin our experimental inquiries here by asking, What is the meaning of blackness? Pass a black ribbon in succession through the colors of the spectrum; it quenches all. This is the meaning of blackness—it is the result of the absorption of *all* the constituents of solar light. Pass a red ribbon through the spectrum. In the red light the ribbon is a vivid red. Why? Because the light that enters the ribbon is not quenched or absorbed, but sent back to the eye. Place the same ribbon in the green or blue of the spectrum; it is black as jet. It absorbs the green and blue light, and leaves the space on which they fall a space of intense darkness. Place a green ribbon in the green of the spectrum. It shines vividly with its proper color; transfer it to the red, it is black as jet. Here it absorbs all the light that falls upon it, and offers mere darkness to the eye. When white light is employed, the red sifts it by quenching the green, and the green sifts it by quenching the red, both exhibiting the residual color. Thus the process through which natural bodies acquire their colors is a *negative* one. The colors are produced by subtraction, not by addition. This red glass is red because it destroys all the more refrangible rays of the spectrum. This blue liquid is blue because it destroys all the less refrangible rays. Both together are opaque because the light transmitted by the one is quenched by the other. In this way by the union of two transparent substances we obtain a combination as dark as pitch to solar light. This other liquid finally is purple because it destroys the green and the yellow, and allows the terminal colors of the spectrum to pass unimpeded. From the blending of the blue and the red this gorgeous color is produced.

These experiments prepare us for the further consideration of a point already adverted to, and regarding which error has found currency for ages. You will find it stated in books that blue and yellow lights mixed together produce green. But blue and yellow have been just proved to be complementary colors, producing white by their mixture. The mixture of blue and yellow *pigments* undoubtedly produces green, but the mixture of pigments is totally different from the mixture of lights. Helmholtz, who first proved

MIXTURE OF PIGMENTS.

yellow and blue to be complementary colors, has revealed the cause of the green in the case of the pigments. No natural color is *pure*. A blue liquid or a blue powder permits not only the blue to pass through it, but a portion of the adjacent green. A yellow powder is transparent not only to the yellow light, but also in part transparent to the adjacent green. Now, when blue and yellow are mixed together, the blue cuts off the yellow, the orange, and the red; the yellow, on the other hand, cuts off the violet, the indigo, and the blue. Green is the only color to which both are transparent, and the consequence is that, when white light falls upon a mixture of yellow and blue powders, the green alone is sent back to the eye. I have already shown you that the fine blue ammonia-sulphate of copper transmits a large portion of green, while cutting off all the less refrangible light. A yellow solution of picric acid also allows the green to pass, but quenches all the more refrangible light. What must occur when we send a beam through both liquids? The green band of the spectrum alone remains upon the screen.

This question of absorption is one of the most subtle and difficult in molecular physics. We are not yet in a condition to grapple with it, but we shall be by-and-by. Meanwhile we may profitably glance back on the web of relations which these experiments reveal to us. We have in the first place in solar light an agent of exceeding complexity, composed of innumerable constituents, refrangible in different degrees. We find, secondly, the atoms and molecules of bodies gifted with the power of sifting solar light in the most various ways, and producing by this sifting the colors observed in nature and art. To do this they must possess a molecular structure commensurate in complexity with that of light itself. Thirdly, we have the human eye and brain so organized as to be able to take in and distinguish the multitude of impressions thus generated. Thus the light at starting is complex; to sift and select it as they do natural bodies must be complex. Finally, to take in the impressions thus generated, the human eye and brain must be highly complex. Whence this triple complexity? If what are called material purposes were the only end to be served, a much simpler mechanism would be sufficient. But, instead of simplicity—instead of the principle of parsimony—we have prodigality of relation and adaptation, and this apparently for the sole purpose of enabling us to see things robed in the splendors of color. Would it not seem that Nature harbored the intention of educating us for other enjoyments than those derivable from meat and drink? At all events, whatever Nature meant—and it would be mere presumption to dogmatize as to what she meant—we find ourselves here as the issue and upshot of her operations, endowed with capacities to enjoy not only the materially useful, but endowed with others of indefinite scope and application, which deal alone with the beautiful and the true.

LECTURE II.

Origin of Physical Theories: Scope of the Imagination: Newton and the Emission Theory: Verification of Physical Theories: The Luminiferous Ether: Wave-Theory of Light: Thomas Young: Fresnel and Arago: Conceptions of Wave-Motion: Interference of Waves: Constitution of Sound-Waves: Analogies of Sound and Light: Illustrations of Wave-Motion: Interference of Sound-Waves: Optical Illustrations: Pitch and Color: Lengths of the Waves of Light and Rates of Vibration of the Ether-Particles: Interference of Light: Phenomena which first suggested the Undulatory Theory: Hooke and the Colors of Thin Plates: The Soap-Bubble: Newton's Rings: Theory of "Fits:" Its Explanation of the Rings: Overthrow of the Theory: Colors of Mother-of-Pearl.

WE might vary and extend our experiments on light indefinitely, and they certainly would prove us to possess a wonderful mastery over the phenomena. But the vesture of the agent only would thus be revealed, not the agent itself. The human mind, however, is so constituted and so educated as regards natural things, that it can never rest satisfied with this outward view of them. Brightness and freshness take possession of the mind when it is crossed by the light of principles, which show the facts of Nature to be organically connected.

Let us, then, inquire what this thing is that we have been generating, reflecting, refracting, and analyzing.

In doing this, we shall learn that the life of the experimental philosopher is twofold. He lives, in his vocation, a life of the senses, using his hands, eyes, and ears in his experiments, but such a question as that now before us carries him beyond the margin of the senses. He cannot consider, much less answer, the question, "What is light?" without transporting himself to a world which underlies the sensible one, and out of which, in accordance with rigid law, all optical phenomena spring. To realize this subsensible world, if I may use the term, the mind must possess a certain pictorial power. It has to visualize the invisible. It must be able to form definite images of the things which that subsensible world contains; and to say that, if such or such a state of things exist in that world, then the phenomena which appear in ours must, of necessity, grow out of this state of things. If the picture be correct, the phenomena are accounted for; a physical theory has been enunciated which unites and explains them all.

This conception of physical theory implies, as you perceive, the exercise of the imagination. Do not be afraid of this word, which seems to render so many respectable people, both in the ranks of science and out of them, uncomfortable. That men in the ranks of science should feel thus is, I think, a proof that they have suffered themselves to be misled by the popular definition of a great faculty instead of observing its operation in their own minds. Without imagination we cannot take a step beyond the bourne of the mere animal world, perhaps not even to the edge of this. But, in speaking thus of imagination, I do not mean a riotous power which deals capriciously with facts, but a well-ordered and disciplined power, whose sole function is to form conceptions which the intellect imperatively demands. Imagination thus exercised never really severs itself from the world of fact. This is the storehouse from which all its pictures are drawn; and the magic of its art consists, not in creating things anew, but in so changing the magnitude, position, and other relations of sensible things, as to render them fit for the requirements of the intellect in the subsensible world.[1]

[1] The following charming extract, bearing upon this point, was discovered and written out for me by my friend Dr. Bence Jones, Hon. Secretary to the Royal Institution:

"In every kind of magnitude there is a degree or sort to which our sense is proportioned, the perception and knowledge of which is of greatest use to mankind. The same is the groundwork of philosophy; for, though all sorts and degrees are equally the object of philosophical speculation, yet it is from those which are proportioned to sense that a philosopher must set out in his inquiries, ascending or descending afterwards as his pursuits may require. He does well indeed to take his views from many points of sight, and supply the defects of sense by a well-regulated imagination; nor is he to be confined by any limit in space or time; but, as his knowledge of Nature is founded on the observation of sensible things, he must begin with these, and must often return to them to examine his progress by them. Here is his secure hold; and as he sets out from thence, so if he likewise trace not often his steps backwards

I will take, as an illustration of this subject, the case of Newton. Before he began to deal with light, he was intimately acquainted with the laws of elastic collision, which all of you have seen more or less perfectly illustrated on a billiard-table. As regards the collision of sensible masses, Newton knew the angle of incidence to be equal to the angle of reflection, and he also knew that experiment, as shown in our last lecture, had established the same law with regard to light. He thus found in his previous knowledge the material for theoretic images. He had only to change the magnitude of conceptions already in his mind to arrive at the Emission Theory of Light. He supposed light to consist of elastic particles of inconceivable minuteness shot out with inconceivable rapidity by luminous bodies. Such particles impinging upon smooth surfaces were reflected in accordance with the ordinary law of elastic collision. The fact of optical reflection certainly occurred as if light consisted of elastic particles, and this was Newton's sole justification for introducing them.

But this is not all. In another important particular, also, Newton's conceptions regarding the nature of light were influenced by his previous knowledge. He had been working at the phenomena of gravitation, and

with caution, he will be in hazard of losing his way in the labyrinths of Nature."—(*Maclaurin: An Account of Sir I. Newton's Philosophical Discoveries. Written* 1728; *second edition,* 1750; pp. 18, 19.)

had made himself at home amid the operations of this universal power. Perhaps his mind at this time was too freshly and too deeply imbued with these notions to permit of his forming an unfettered judgment regarding the nature of light. Be that as it may, Newton saw in *refraction* the action of an attractive force exerted on the light-particles. He carried his conception out with the most severe consistency. Dropping vertically downwards towards the earth's surface, the motion of a body is accelerated as it approaches the earth. Dropping in the same manner downwards on a horizontal surface, say through air on glass or water, the velocity of the light-particles, when they come close to the surface, was, according to Newton, also accelerated. Approaching such a surface obliquely, he supposed the particles, when close to it, to be drawn down upon it, as a projectile is drawn by gravity to the surface of the earth. This deflection was, according to Newton, the refraction seen in our last lecture. Finally, it was supposed that differences of color might be due to differences in the sizes of the particles. This was the physical theory of light enunciated and defended by Newton; and you will observe that it simply consists in the transference of conceptions born in the world of the senses to a subsensible world.

But, though the region of physical theory lies thus behind the world of senses, the verifications of theory occur in that world. Laying the theoretic conception at the root of matters, we determine by rigid deduction what are the phenomena which must of necessity grow out of this root. If the phenomena thus deduced agree with those of the actual world, it is a presumption in favor of the theory. If as new classes of phenomena arise they also are found to harmonize with theoretic deduction, the presumption becomes still stronger. If, finally, the theory confers prophetic vision upon the investigator, enabling him to predict the existence of phenomena which have never yet been seen, and if those predictions be found on trial to be rigidly correct, the persuasion of the truth of the theory becomes overpowering. Thus working backwards from a limited number of phenomena, genius, by its own expansive force, reaches a conception which covers all the phenomena. There is no more wonderful performance of the intellect than this. And we can render no account of it. Like the scriptural gift of the Spirit, no man can tell whence it cometh. The passage from fact to principle is sometimes slow, sometimes rapid, and at all times a source of intellectual joy. When rapid, the pleasure is concentrated and becomes a kind of ecstasy or intoxication. To any one who has experienced this pleasure, even in a moderate degree, the action of Archimedes when he quitted the bath, and ran naked, crying "Eureka!" through the streets of Syracuse, becomes intelligible.

How, then, did it fare with the theory of Newton,

when the deductions from it were brought face to face with natural phenomena? To the mind's eye, Newton's elastic particles present themselves like particles of sensible magnitude. The same reasoning applies to both; the same experimental checks exist for both. Tested by experiment, then, Newton's theory was found competent to explain many facts, and with transcendent ingenuity its author sought to make it account for all. He so far succeeded, that men so celebrated as Laplace and Malus, who lived till 1812, and Biot and Brewster, who lived till our own time, were found among his disciples.

Still, even at an early period of the existence of the Emission Theory, one or two great names were found recording a protest against it; and they furnish another illustration of the law that, in forming theories, the scientific imagination must draw its materials from the world of fact and experience. It was known long ago that sound is conveyed in waves or pulses through the air; and no sooner was this truth well housed in the mind than it was transformed into a theoretic conception. It was supposed that light, like sound, might also be the product of wave-motion. But what, in this case, could be the material forming the waves? For the waves of sound we have the air of our atmosphere; but the stretch of imagination which filled all space with a *luminiferous ether* trembling with the waves of light was so bold as to shock cautious minds. In one of my latest conversations with Sir David Brewster he said to me that his chief objection to the undulatory theory of light was that he could not think the Creator guilty of so clumsy a contrivance as the filling of space with ether in order to produce light. This, I may say, is very dangerous ground, and the quarrel of science with Sir David, on this point, as with many other persons on other points, is, that they profess to know too much about the mind of the Creator.

This conception of an ether was advocated and indeed applied to various phenomena of optics by the celebrated astronomer, Huyghens. It was espoused and defended by the celebrated mathematician, Euler. They were, however, opposed by Newton, whose authority at the time bore them down. Or shall I say it was authority merely? Not quite so. Newton's preponderance was in some degree due to the fact that, though Huyghens and Euler were right in the main, they did not possess sufficient data to *prove* themselves right. No human authority, however high, can maintain itself against the voice of Nature speaking through experiment. But the voice of Nature may be an uncertain voice, through the scantiness of data. This was the case at the period **now** referred to, and at such a period by the authority of Newton all antagonists were naturally overborne.

Still, this great **Emission** Theory, which held its ground so long, resembled one of those circles which, according to your countryman Emerson, the force of genius periodically draws round the operations of the intellect, but which are eventually broken through by pressure from behind. In the year 1773 was born, at Milverton, in Somersetshire, one of the most remarkable men that England ever produced. He was educated for the profession of a physician, but was too strong to be tied down to professional routine. He devoted himself to the study of natural philosophy, and became in all its departments a master. He was also a master of letters. Languages, ancient and modern, were housed within his brain, and, to use the words of his epitaph, "he first penetrated the obscurity which had veiled for ages the hieroglyphics of Egypt." It fell to the lot of this man to discover facts in optics which Newton's theory was incompetent to explain, and his mind roamed in search of a sufficient theory. He had made himself acquainted with all the phenomena of wave-motion; with all the phenomena of sound; working successfully in this domain as an original discoverer. Thus informed and disciplined, he was prepared to detect any resemblance which might reveal itself between the phenomena of light and those of wave-motion. Such resemblances he did detect; and, spurred on by the discovery, he pursued his speculations and his experiments, until he finally succeeded in placing on an immovable basis the Undulatory Theory of Light.

The founder of this great theory was Thomas Young, a name, perhaps, unfamiliar to many of you.

Permit me, by a kind of geometrical construction which I once employed in London, to give you a notion of the magnitude of this man. Let Newton stand erect in his age, and Young in his. Draw a straight line from Newton to Young, which shall form a tangent to the heads of both. This line would slope downwards from Newton to Young, because Newton was certainly the taller man of the two. But the slope would not be steep, for the difference of stature was not excessive. The line would form what engineers call a gentle gradient from Newton to Young. Place underneath this line the biggest man born in the interval between both. He would not, in my opinion, reach the line; for if he did he would be taller intellectually than Young, and there was, I believe, none taller. But I do not want you to rest on English estimates of Young; the German, Helmholtz, a kindred genius, thus speaks of him: "His was one of the most profound minds that the world has ever seen; but he had the misfortune to be too much in advance of his age. He excited the wonder of his contemporaries, who, however, were unable to follow him to the heights at which his daring intellect was accustomed to soar. His most important ideas lay, therefore, buried and forgotten in the folios of the Royal Society, until a new generation gradually and painfully made the same discoveries, and proved the exactness of his assertions and the truth of his demonstrations."

It is quite true, as Helmholtz says, that Young was in advance of his age; but something is to be added which illustrates the responsibility of our public writers. For twenty years this man of genius was quenched—hidden from the appreciative intellect of his countrymen—deemed in fact a dreamer, through the vigorous audacity of a writer who had then possession of the public ear, and who in the *Edinburgh Review* poured ridicule upon Young and his speculations. To the celebrated Frenchmen Fresnel and Arago, he was first indebted for the restitution of his rights, for they, especially Fresnel, remade independently, as Helmholtz says, and vastly extended his discoveries. To the students of his works Young has long since appeared in his true light, but these twenty blank years pushed him from the public mind, which became in turn filled with the fame of Young's colleague at the Royal Institution, Davy, and afterwards with the fame of Faraday. Carlyle refers to a remark of Novalis, that a man's self-trust is enormously increased the moment he finds that others believe in him. If the opposite remark be true—if it be a fact that public disbelief weakens a man's force—there is no calculating the amount of damage these twenty years of neglect may have done to Young's productiveness as an investigator. It remains to be stated that his assailant was Mr. Henry Brougham, afterwards Lord Chancellor of England.

Our hardest work is now before us. And, as I have often had occasion to notice that capacity for hard work depends in a great measure on the antecedent winding up of the will and determination, I would call upon you to gird up your loins for our coming labors. If we succeed in climbing the hill which faces us to-night, our future efforts will be comparatively light.

In the earliest writings of the ancients we find the notion that sound is conveyed by the air. Aristotle gives expression to this notion, and the great architect Vitruvius compares the waves of sound to waves of water. But the real mechanism of wave-motion was hidden from the ancients, and indeed was not made clear until the time of Newton. The central difficulty of the subject was, to distinguish between the motion of the wave itself and the motion of the particles which at any moment constitute the wave.

Stand upon the sea-shore and observe the advancing rollers before they are distorted by the friction of the bottom. Every wave has a back and a front, and, if you clearly seize the image of the moving wave, you will see that every particle of water along the front of the wave is in the act of rising, while every particle along its back is in the act of sinking. The particles in front reach in succession the crest of the wave, and as soon as the crest is passed they begin to fall. They then reach the furrow or *sinus* of the wave, and can sink no farther. Immediately afterwards they become the front of the succeeding wave, rise again until they reach the crest, and then sink as before. Thus, while the waves pass onward horizontally, the individual particles are simply lifted up and down vertically. Observe a sea-fowl, or, if you are a swimmer, abandon yourself to the action of the waves; you are not carried forward, but simply rocked up and down. The propagation of a wave is the propagation of *a form*, and not the transference of the substance which constitutes the wave.

The *length* of the wave is the distance from crest to crest, while the distance through which the individual particles oscillate is called the *amplitude* of the oscillation. You will notice that in this description the particles of water are made to vibrate *across* the line of propagation.[1]

And now we have to take a step forward, and it is the most important step of all. You can picture two series of waves proceeding from different origins through the same water. When, for example, you throw two stones into still water, the ring-waves proceeding from the two centres of disturbance intersect each other. Now, no matter how numerous these waves may be, the law holds good that the motion of every particle of the water is the algebraic sum of all the mo-

[1] I do not wish to encumber the conception here with the details of the motion, but I may draw attention to the beautiful model of Prof. Lyman, wherein waves are shown to be produced by the *circular* motion of the particles. This, as proved by the brothers Weber, is the real motion in the case of water-waves.

tions imparted to it. If crest coincide with crest, the wave is lifted to a double height; if furrow coincide with crest, the motions are in opposition, and their sum is zero. We have then still water, which we shall learn presently corresponds to what we call *darkness* in reference to our present subject. This action of wave upon wave is technically called *interference*, a term to be remembered.

Thomas Young's fundamental discovery in optics was that the principle of Interference applied to light. Long prior to his time an Italian philosopher, Grimaldi, had stated that under certain circumstances two thin beams of light, each of which, acting singly, produced a luminous spot upon a white wall, when caused to act together, partially quenched each other and darkened the spot. This was a statement of fundamental significance, but it required the discoveries and the genius of Young to give it meaning. How he did so, I will now try to make clear to you. You know that air is compressible; that by pressure it can be rendered more dense, and that by dilatation it can be rendered more rare. Properly agitated, a tuning-fork now sounds in a manner audible to you all, and most of you know that the air through which the sound is passing is parcelled out into spaces in which the air is condensed, followed by other spaces in which the air is rarefied. These condensations and rarefactions constitute what we call *waves* of sound. You can imagine the air of a room traversed by a series of such waves, and you can imagine a second series sent through the same air, and so related to the first that condensation coincides with condensation and rarefaction with rarefaction. The consequence of this coincidence would be a louder sound than that produced by either system of waves taken singly. But you can also imagine a state of things where the condensations of the one system fall upon the rarefactions of the other system. In this case the two systems would completely neutralize each other. Each of them taken singly produces sound; both of them taken together produce no sound. Thus, by adding sound to sound we produce silence, as Grimaldi in his experiment produced darkness by adding light to light.

The analogy between sound and light here at once flashes upon the mind. Young generalized this observation. He discovered a multitude of similar cases, and determined their precise conditions. On the assumption that light was wave-motion, all his experiments on interference were explained; on the assumption that light was flying particles, nothing was explained. In the time of Huyghens and Euler a medium had been assumed for the transmission of the waves of light; but Newton raised the objection that, if light consisted of the waves of such a medium, shadows could not exist. The waves, he contended, would bend round opaque bodies and produce the motion of light behind them, as sound turns a corner, or as waves of water wash round a rock. It was proved that the bending round referred to by Newton actually occurs, but that the inflected waves abolish each other by their mutual interference. Young also discerned a fundamental difference between the waves of light and those of sound. Could you see the air through which sound-waves are passing, you would observe every individual particle of air oscillating to and fro in the direction of propagation. Could you see the ether, you would also find every individual particle making a small excursion to and fro, but here the motion, like that assigned to the water-particles above referred to, would be *across* the line of propagation. The vibrations of the air are *longitudinal*, the vibrations of the ether are *transversal*.

It is my desire that you should realize with clearness the character of wave-motion, both in ether and in air. And, with this view, I bring before you an experiment wherein the air-particles are represented by small spots of light. They are parts of a spiral, drawn upon a circle of blackened glass, and, when the circle rotates, the spots move in successive pulses over the screen. You have here clearly set before you how the pulses travel incessantly forward, while the particles that compose them perform oscillations to and fro. This is the picture of a sound-wave, in which the vibrations are longitudinal. By another glass wheel, we produce an image of a transverse wave, and here we observe the waves travelling in succession over the screen, while each individual spot of light performs an excursion to and fro across the line of propagation.

Notice what follows when the glass wheel is turned very quickly. Objectively considered, the transverse waves propagate themselves as before, but subjectively the effect is totally changed. Because of the retention of impressions upon the retina, the spots of light simply describe a series of parallel luminous lines upon the screen, the length of these lines marking the amplitude of the vibration. The impression of wave-motion has totally disappeared.

The most familiar illustration of the interference of sound-waves is furnished by the *beats* produced by two musical sounds slightly out of unison. These two tuning-forks are now in perfect unison, and when they are agitated together the two sounds flow without roughness, as if they were but one. But, by attaching to one of the forks a two-cent piece, we cause it to vibrate a little more slowly than its neighbor. Suppose that one of them performs 101 vibrations in the time required by the other to perform 100, and suppose that at starting the condensations and rarefactions of both forks coincide. At the 101st vibration of the quickest fork they will again coincide, the quicker fork at this point having gained one whole vibration, or one whole wave upon the other. But a little reflection will make it clear that, at the 50th vibration, the two forks are in opposition; here the one tends to produce a condensation where the other tends to produce a rarefaction; by the united action of the two forks, therefore, the sound is quenched, and we have a pause of silence. This occurs where one fork has gained *half a wavelength* upon the other. At the 101st vibration we have again coincidence, and, therefore, augmented sound; at the 150th vibration we have again a quenching of the sound. Here the one fork is *three half-waves* in advance of the other. In general terms, the waves conspire when the one series is an *even* number of half-wave lengths, and they are destroyed when the one series is an *odd* number of half-wave lengths in advance of the other. With two forks so circumstanced, we obtain those intermittent shocks of sound separated by pauses of silence, to which we give the name of beats.

I now wish to show you what may be called the *optical expression* of those beats. Attached to a large tuning-fork, F (Fig. 2), is a small mirror, which shares the vibrations of the fork, and on to the mirror is thrown a thin beam of light, which shares the vibrations of the mirror. The beam reflected from the fork is received upon a piece of looking-glass, and thrown back upon the screen, where it stamps itself as a small luminous disk. The agitation of the fork by a violin-bow converts that disk into a *band* of light, and if you simply move your heads to and fro you cause the image of the band to sweep over the retina, drawing it out to a sinuous line, thus proving the periodic character of the motion which produces it. By a sweep of the looking-glass we can also cover the screen from side to side by a luminous scroll, *m n*, Fig. 2, the depth of the sinuosities indicating the amplitude of the vibration.

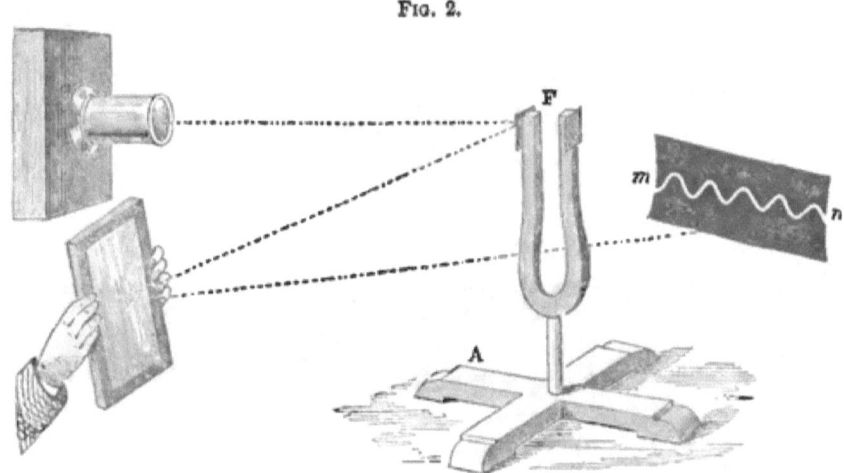

Fig. 2.

Instead of receiving the beam reflected from the fork on a piece of looking-glass, we now receive it upon a second mirror attached to a second fork, and cast by it upon the screen. Both forks now act in combination upon the beam. The disk is drawn out, as before, the band of light gradually shortening as the motion subsides, until, when the motion ceases, we have our luminous disk restored. Weighting one of the forks as we did before, with a two-cent piece, sometimes the forks conspire, and then you have the band of light drawn out to its maximum length; sometimes they oppose each other, and then you have the band of light diminished to a circle. Thus, the beats which address the ear express themselves optically as the alternate elongation and shortening of the band of light. If I move the mirror of this second fork, you have a sinuous line, as before; but the sinuosities are sometimes deep, and sometimes they almost disappear, as in Fig. 3, thus expressing the alternate increase and diminution of the sound, the intensity of which is expressed by the depth of the sinuosities. To Lissajous we owe this mode of illustration.

Fig. 8.

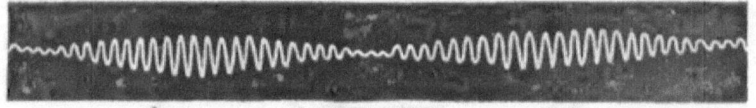

The *pitch* of a sound is wholly determined by the rapidity of the vibration, as the *intensity* is by the amplitude. The rise of pitch with the rapidity of the impulses may be illustrated by the syren, which consists of a perforated disk rotating over a cylinder into which air is forced, and the end of which is also perforated. When the perforations of the disk coincide with those of the cylinder, a puff escapes; and, when the puffs succeed each other with sufficient rapidity, the impressions upon the auditory nerve link themselves together to a continuous musical note. The more rapid the rotation of the disk the quicker is the succession of the impulses, and the higher the pitch of the note. Indeed, by means of the syren the number of vibrations due to any musical note, whether it be that of an instrument, of the human voice, or of a flying insect, may be accurately determined.

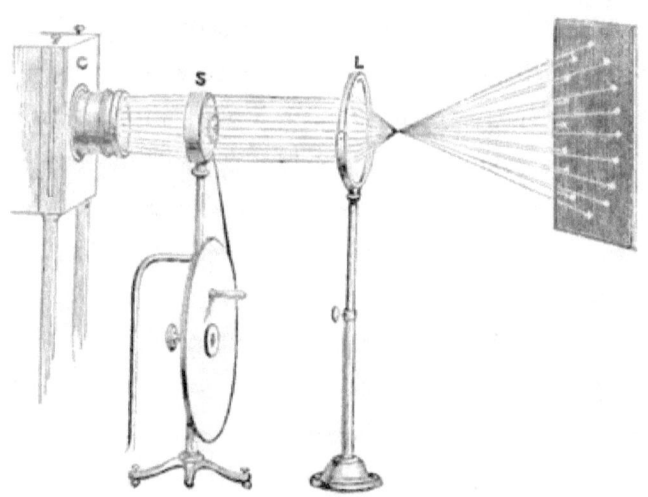

Fig. 4.

In front of our lamp now stands a very homely instrument S, Fig. 4, of this character. The perforated disk is turned by a wheel and band, and, when the two sets of perforations coincide, a series of spots of light, sharply defined by the lens L, ranged on the circumference of a circle, is seen upon the screen. On slowly turning the disk, a flicker is produced by the alternate stoppage and transmission of the light. At the same time air is urged into the syren, and you hear a fluttering sound corresponding to the flickering light. But, by augmenting the rapidity of rotation, the light, though intercepted as before, appears perfectly steady, through the persistence of impressions upon the retina; and, about the time when the optical impression becomes continuous, the auditory impression becomes equally so; the puffs from the syren linking themselves then together to a continuous musical note, which rises in pitch with the rapidity of the rotation. A movement of the head causes the image of the spots to sweep over the retina, producing beaded lines: the same effect is produced upon our screen by the sweep of a looking-glass which has received the thin beams from the syren.

In the undulatory theory, what pitch is to the ear, color is to the eye. Though never seen, the lengths of the waves of light have been determined. Their existence is proved *by their effects*, and from their effects also their lengths may be accurately deduced. This may, moreover, be done in many ways, and, when the different determinations are compared, the strictest harmony is found to exist between them. The shortest waves of the visible spectrum are those of the extreme violet; the longest, those of the extreme red; while the other colors are of intermediate pitch or wave-length. The length of a wave of the extreme red is such that it would require 36,918 of them placed end to end to cover one inch, while 64,631 of the extreme violet waves would be required to span the same distance.

Now, the velocity of light, in round numbers, is 190,000 miles per second. Reducing this to inches, and multiplying the number thus found by 36,918, we obtain the number of waves of the extreme red in 190,000 miles. *All these waves enter the eye, and hit the retina at the back of the eye in one second.* The number of shocks per second necessary to the production of the impression of red is, therefore, four hundred and fifty-one millions of millions. In a similar manner, it may be found that the number of shocks corresponding to the impression of violet is seven hundred and eighty-nine millions of millions. All space is filled with matter oscillating at such rates. From every star waves of these dimensions move with the velocity of light like spherical shells outwards. And in the ether, just as in the water, the motion of every particle is the algebraic sum of all the separate motions imparted to it. Still, one motion does not blot the other out; or, if extinction occur at one point, it is atoned for at some other point. Every star declares by its light its undamaged individuality, as if it alone had sent its thrills through space.

The principle of interference applies to the waves of light as it does to the waves of water and the waves of sound. And the conditions of interference are the same in all three. If two series of light-waves of the same length start at the same moment from a common origin, crest coincides with crest, sinus with sinus, and the two systems blend together to a single system of double amplitude. If both series start at the same moment, one of them being, at starting, a whole wavelength in advance of the other, they also add themselves together, and we have an augmented luminous effect. Just as in the case of sound, the same occurs when the one system of waves is any *even* number of semi-undulations in advance of the other. But if the one system be half a wave-length, or any *odd* number of half wave-lengths in advance, then the crests of the one fall upon the sinuses of the other; the one system, in fact, tends to *lift* the particles of ether at the precise places where the other tends to *depress* them; hence, through their joint action the ether remains perfectly still. This stillness of the ether is what we call darkness, which corresponds, as already stated, with a dead level in the case of water.

It was said in our first lecture, with reference to the colors produced by absorption, that the function of natural bodies is selective, not creative; that they extinguish certain constituents of the white solar light, and appear in the colors of the unextinguished light. It must at once flash upon your minds that, inasmuch as we have in interference an agency by which light may be self-extinguished, we may have in it the conditions for the production of color. But this would imply that certain constituents are quenched by interference, while others are permitted to remain. This is the fact; and it is entirely due to the difference in the lengths of the waves of light.

The subject is most easily illustrated by the class of phenomena which first suggested the undulatory theory to the mind of Hooke. These are the colors of thin films of all kinds, which are known as the *colors of thin plates.* In this relation no object in the world possesses a deeper scientific interest than a common soap-bubble. And here let me say emerges one of the difficulties which the student of pure science encounters in the presence of "practical" communities like those of America and England; it is not to be expected that such communities can entertain any profound sympathy with labors which seem so far removed from the domain of practice as many of the labors of the man of science are. Imagine Dr. Draper spending his days in blowing soap-bubbles and in studying their colors! Would you show him the necessary patience, or grant him the necessary support? And yet be it remembered it was thus that Newton spent a large portion of his time; and that on such experiments has been founded a theory, the issues of which are incalculable. I see no other way for you laymen than to trust the scientific man with the choice of his inquiries; he stands before the tribunal of his peers, and by their verdict on his labors you ought to abide.

Whence, then, are derived the colors of the soap-bubble? Imagine a beam of white light impinging on the bubble. When it reaches the first surface of the film, a known fraction of the light is reflected back. But a large portion of the beam enters the film, reaches its second surface, and is again in part reflected. The waves from the second surface thus turn back and hotly pursue the waves from the first surface. And, if the thickness of the film be such as to cause the necessary retardation, the two systems of waves interfere with each other, producing augmented or diminished light, as the case may be. But, inasmuch as the waves of light are of different lengths, it is plain that, to produce self-extinction in the case of the longer waves, a greater thickness of film is necessary than in the case of the shorter ones. Different colors, therefore, appear at different thicknesses of the film.

Take with you a little bottle of spirit of turpentine, and pour it into one of the ponds in the Central Park. You will then see the flashing of those colors over the surface of the water. On a small scale we produce them thus: A common tea-tray is filled with water, beneath the surface of which dips the end of a pipette. A beam of light falls upon the water, and is reflected by it to the screen. Spirit of turpentine is poured into the pipette; it descends, issues from the end in minute drops, which rise in succession to the surface. On reaching it, each drop spreads suddenly out as a film, and glowing colors immediately flash forth upon the screen. The colors change as the thickness of the film changes by evaporation. They are also arranged in zones in consequence of the gradual diminution of thickness from the centre outwards.

Any film whatever will produce these colors. The film of air between two plates of window-glass, squeezed together, exhibits rich fringes of color. Nor is even air necessary; the mere rupture of optical continuity suffices. Smite with an axe the black, transparent ice—black, because it is transparent and of great depth—under the moraine of a glacier; you readily produce in the interior flaws which no air can reach, and from these flaws the colors of thin plates sometimes break like fire. The colors are commonly seen in flawed crystals; they are also formed by the film of oxide which collects upon molten lead. It is the colors of thin plates that guide the tempering of steel. But the origin of most historic interest is, as already stated, the soap-bubble. With one of these mixtures employed by the eminent blind philosopher Plateau in his researches on the cohesion figures of thin films, we obtain in still air a bubble twelve or fifteen inches in diameter. You may look at the bubble itself, or you may look at its projection upon the screen, rich colors arranged in zones are, in both cases, exhibited. Rendering the beam parallel, and permitting it to impinge upon the sides, bottom, and top, of the bubble, gorgeous fans of color overspread the screen, which rotate as the beam is carried round the circumference of the bubble. By this experiment the internal motions of the film are also strikingly displayed.

Newton sought to measure the thickness of the bubble corresponding to each of these colors; in fact, he sought to determine generally the relation of color to thickness. His first care was to obtain a film of variable and calculable depth. On a plano-convex glass lens of very feeble curvature he laid a plate of glass with a plane surface, thus obtaining a film of air of gradually increasing depth from the point of contact outwards. On looking at the film in monochromatic light he saw surrounding the place of contact a series of bright rings separated from each other by dark ones, and becoming more closely packed together as the distance from the point of contact augmented. When he employed *red* light, his rings had certain diameters; when he employed *blue* light, the diameters were less. Causing his glasses to pass through the spectrum from red to blue, the rings contracted; when the passage was from blue to red, the rings expanded. When *white* light fell upon the glasses, inasmuch as the colors were not superposed, a series of *iris-colored* circles were obtained. They became paler as the film became thicker, until finally the colors became so intimately reblended as to produce white light. A magnified image of *Newton's rings* is now before you, and, by employing in succession red, blue, and white light, we obtain all the effects observed by Newton.

He compared the tints thus obtained with the tints of his soap-bubble, and he calculated the corresponding thickness. How he did this may be thus made plain to you: Suppose the water of the ocean to be absolutely smooth; it would then accurately represent the earth's curved surface. Let a perfectly horizontal plane touch the surface at any point. Knowing the earth's diameter, any engineer or mathematician in this room could tell you how far the sea's surface will lie below this plane, at the distance of a yard, ten yards, a hundred yards, or a thousand yards from the point of contact of the plane and the sea. It is common, indeed, in levelling operations, to allow for the curvature of the earth. Newton's calculation was precisely similar. His plane glass was a tangent to his curved one. From its refractive index and focal distance he determined the diameter of the sphere of which his curved glass formed a segment, he measured the distances of his rings from the place of contact, and he calculated the depth between the tangent plane and the curved surface, exactly as the engineer would calculate the distance between his tangent plane and the surface of the sea. The wonder is, that, where such infinitesimal distances are involved, Newton, with the means at his disposal, could have worked with such marvellous exactitude.

To account for these rings was the greatest difficulty that Newton ever encountered. He quite appreciated the difficulty. Over his eagle-eye there was no film—no vagueness in his conceptions. At the very outset his theory was confronted by the question, Why, when a beam of light is incident on a transparent body, are some of the light-particles reflected and some transmitted? Is it that there are two kinds of particles, the one specially fitted for transmission and the other for reflection? This cannot be the reason; for, if we allow a beam of light which has been reflected from one piece of glass to fall upon another, it, as a general rule, is also divided into a reflected and a transmitted portion. Thus the particles once reflected are not always reflected, nor are the particles once transmitted always transmitted. Newton saw all this; he knew he had to explain why it is that the self-same particle is at one moment reflected and at the next moment transmitted. It could only be through *some change in the condition of the particle itself*. The self-same particle, he affirmed, was affected by "fits" of easy transmission and reflection.

If you are willing to follow me while I unravel this theory of fits, the most subtle, perhaps, that ever entered the human mind, the intellectual discipline will repay you for the necessary effort of attention. Newton was chary of stating what he considered to be the cause of the fits, but there cannot be a doubt that his

THEORY OF FITS.

mind rested on a mechanical cause. Nor can there be a doubt that, as in all attempts at theorizing, he was compelled to fall back upon experience for the materials of his theory. His course of observation and of thought may have been this: From a magnet he might obtain the notion of attracted and repelled poles. What more natural than that he should endow his light-particles with such poles? Turning their attracted poles towards a transparent substance, the particles would be sucked in and transmitted; turning their repelled poles, they would be driven away or reflected. Thus, by the ascription of poles, the transmission and reflection of the self-same particle at different times might be accounted for.

Regard these rings of Newton as seen in pure red light: they are alternately bright and dark. The film of air corresponding to the outermost of them is not thicker than an ordinary soap-bubble, and it becomes thinner on approaching the centre; still Newton, as I have said, measured the thickness corresponding to every ring and showed the difference of thickness between ring and ring. Now, mark the result. For the sake of convenience, let us call the thickness of the film of air corresponding to the first dark ring d; then Newton found the distance corresponding to the second dark ring $2d$; the thickness corresponding to the third dark ring $3d$; the thickness corresponding to the tenth dark ring $10d$, and so on.

Surely there must be some hidden meaning in this little distance *d*, which turns up so constantly? One can imagine the intense interest with which Newton pondered its meaning. Observe the outcome of his thought. He had probably endowed his light-particles with poles, but now he is forced to introduce the notion of *periodic recurrence.* How was this to be done? By supposing the light-particles animated, not only with a motion of translation, but also with a motion of rotation. Newton's astronomical knowledge would render all such conceptions familiar to him. The earth has such a motion. In the time occupied in passing over a million and a half of miles of its orbit— that is, in twenty-four hours—our planet performs a complete rotation, and, in the time required to pass over the distance *d*, Newton's light-particle must be supposed to perform a complete rotation. True, the light-particle is smaller than the planet, and the distance *d*, instead of being a million and a half of miles, is a little over the ninety thousandth of an inch. But the two conceptions are, in point of intellectual quality, identical.

Imagine, then, a particle entering the film of air where it possesses this precise thickness. To enter the film, its attracted end must be presented. Within the film it is able to turn *once* completely round; at the other side of the film its attracted pole will be again presented; it will, therefore, enter the glass at the opposite side of the film *and be lost to the eye.* All round the place of contact, wherever the film possesses this precise thickness, the light will equally disappear—we shall have a ring of darkness.

And now observe how well this conception falls in with the law of proportionality discovered by Newton. When the thickness of the film is $2d$, the particle has time to perform *two* complete somersaults within the film; when the thickness is $3d$, *three* complete somersaults; when $10d$, *ten* complete somersaults are performed. It is manifest that in each of these cases, on arriving at the second surface of the film, the attracted pole of the particle will be presented. It will, therefore, be transmitted, and, because no light is sent to the eye, we shall have a ring of darkness at each of these places.

The bright rings follow immediately from the same conception. They occur between the dark rings, the thicknesses to which they correspond being also intermediate between those of the dark ones. Take the case of the first bright ring. The thickness of the film is $\frac{1}{2}d$; in this interval the rotating particle can perform only half a rotation. When, therefore, it reaches the second surface of the film, its repelled pole is presented; it is, therefore, driven back and reaches the eye. At all distances round the centre corresponding to this thickness the same effect is produced, and the consequence is a ring of brightness. The other bright rings are similarly accounted for. At the second one, where the thickness is $1\frac{1}{2}d$, a rotation and a half is performed; at the third, two rotations and a half; and at each of these places the particles present their repelled poles to the lower surface of the film. They are therefore sent back to the eye, producing the impression of brightness. Here, then, we have unravelled the most subtle application that Newton ever made of the Emission Theory.

It has been stated in the early part of this lecture, that the Emission Theory assigned a greater velocity to light in glass and water, than in air or stellar space. Here it was at direct issue with the theory of undulation, which makes the velocity in air or stellar space *less* than in glass, or water. By an experiment proposed by Arago, and executed with consummate skill by Foucault and Fizeau, this question was brought to a crucial test, and decided in favor of the theory of undulation. In the present instance also the two theories are at variance. Newton assumed that the action which produces the alternate bright and dark rings took place at a *single surface;* that is, the second surface of the film. The undulatory theory affirms that the rings are caused by the interference of waves reflected from *both surfaces.* This also has been demonstrated by experiment. By proper devices we may abolish reflection from one of the surfaces of the film, and when this is done the rings vanish altogether.

Rings of feeble intensity are also formed by *trans-*

mitted light. These are referred by the undulatory theory to the interference of waves which have passed *directly* through the film, with others which have suffered *two reflections* within the film. They are thus completely accounted for.

Newton, by the foregoing exceedingly subtle assumption, vaulted over the difficulty presented by the colors of thin plates. And, as further difficulties in process of time thickened round the theory, his disciples tried to sustain it with an ingenuity worthy of their master. The new difficulties were not anticipated by the theory, but were met by new assumptions, until at length the Emission Theory became what a distinguished writer calls a "mob of hypotheses." In the presence of the phenomena of interference the theory finally broke down, while the whole of these phenomena lie as it were latent in the theory of undulation. Newton's "fits," for example, are immediately translatable into the lengths of the ether-waves. We have the observed periodic recurrence as the thickness varies so as to produce a retardation of an odd or even number of semi-undulations.[1]

---

[1] In the explanation of Newton's rings, something besides thickness is to be taken into account. In the case of the first surface of the film of air, the waves pass from a denser to a rarer medium, while in the case of the second surface the waves pass from a rarer to a denser medium. This difference at the two reflecting surfaces can be proved to be equivalent *to the addition of half a wave-length* to the thickness of the film. To the absolute

Numerous other colors are due to interference. Fine scratches drawn upon glass or polished metal reflect the waves of light from their sides; and some, being reflected from opposite sides of the same furrow, interfere with and quench each other. But the obliquity of reflection which extinguishes the shorter waves does not extinguish the longer ones, hence the phenomena of color. These are called the colors of *striated surfaces*. They are well illustrated by mother-of-pearl. This shell is composed of exceedingly thin layers, which, when cut across by the polishing of the shell, expose their edges and furnish the necessary small and regular grooves. The most conclusive proof that the colors are due to the mechanical state of the surface is to be found in the fact, established by Brewster, that, by stamping the shell carefully upon black sealing-wax, we transfer the grooves, and produce upon the wax the colors of mother-of-pearl.

thickness as determined by Newton, half a wave-length is in each case to be added. When this is done, the dark and bright rings follow each other in exact accordance with the law of interference already enunciated.

LECTURE III.

Relation of Theories to Experience: Origin of the Notion of the Attraction of Gravitation: Notion of Polarity, how generated: Atomic Polarity: Structural Arrangements due to Polarity: Architecture of Crystals considered as an Introduction to their Action upon Light: Notion of Atomic Polarity applied to Crystalline Structure: Experimental Illustrations: Crystallization of Water: Expansion by Heat and by Cold: Deportment of Water considered and explained: Molecular Action illustrated by a Model: Force of Solidification: Bearings of Crystallization on Optical Phenomena: Refraction: Double Refraction: Polarization: Action of Tourmaline: Character of the Beams emergent from Iceland Spar: Polarization by ordinary Refraction and Reflection: Depolarization.

In our last lecture we sought to familiarize our minds with the characteristics of wave-motion. We drew a clear distinction between the motion of the wave itself and the motion of its constituent particles. Passing through water-waves and air-waves, we prepared our minds for the conception of light-waves propagated through the luminiferous ether. The analogy of sound will fix the whole mechanism in your minds. Here we have a vibrating body which originates the wave-motion, we have, in the air, a vehicle which conveys it, and we have the auditory nerve which receives the impressions of the sonorous waves. In the case of light we have in the vibrating atoms of the luminous body the originators of the wave-motion, we have in the ether its vehicle, while the optic nerve receives the impression of the luminiferous waves. We learned, also, that color is the analogue of pitch, that the rapidity of atomic vibration augmented, and the length of the ether-waves decreased, in passing from the red to the blue end of the spectrum. The fruitful principle of interference we also found applicable to the phenomena of light; and we learned that, in consequence of the different lengths of the ether-waves, they were extinguished by different thicknesses of a transparent film, the particular thickness which quenched one color glowing, therefore, with the complementary one. Thus the colors of thin plates were accounted for.

But one of the objects of our last lecture, and that not the least important, was to illustrate the manner in which scientific theories are formed. They, in the first place, take their rise in the desire of the mind to penetrate to the sources of phenomena. This desire has long been a part of human nature. It prompted Cæsar to say that he would exchange his victories for a glimpse of the sources of the Nile; it may be seen working in Lucretius; it impels Darwin to those daring speculations which of late years have so agitated the public mind. We have learned that in framing theories the imagination does not create, but that it expands, diminishes, moulds, and refines, as the case may be, materials derived from the world of fact and observation.

This is more evidently the case in a theory like that

of light, where the motions of a subsensible medium, the ether, are presented to the mind. But no theory escapes the condition. Newton took care not to encumber gravitation with unnecessary physical conceptions; but we have reason to know that he indulged in them, though he did not connect them with his theory. But even the theory as it stands did not enter the mind as a revelation dissevered from the world of experience. The germ of the conception that the sun and planets are held together by a force of attraction is to be found in the fact that a magnet had been previously seen to attract iron. The notion of matter attracting matter came thus from without, not from within. In our present lecture the magnetic force must serve us still further; but here we must master its elementary phenomena.

The general facts of magnetism are most simply illustrated by a magnetized bar of steel, commonly called a bar magnet. Placing such a magnet upright upon a table, and bringing a magnetic needle near its bottom, one end of the needle promptly retreats from the magnet, while the other as promptly approaches. The needle is held quivering there by some invisible influence exerted upon it. Raising the needle along the magnet, but still avoiding contact, the rapidity of its oscillations decreases, because the force acting upon it becomes weaker. At the centre the oscillations cease. Above the centre, the end of the needle which had been previously drawn towards the magnet retreats, and the opposite end approaches. As we ascend higher, the oscillations become more violent, because the force becomes stronger. At the upper end of the magnet, as at the lower, the force reaches a maximum, but all the lower half of the magnet, from E to S (Fig. 5),

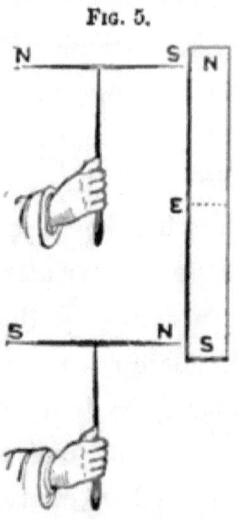

Fig. 5.

attracts one end of the needle, while all the upper half, from E to N, attracts the opposite end. This *doubleness* of the magnetic force is called *polarity*, and the points near the ends of the magnet in which the forces seem concentrated are called its *poles*.

What, then, will occur if we break this magnet in two at the centre E? Will each of the separate halves act as it did when it formed part of the whole magnet? No; each half is in itself a perfect magnet, possessing two poles. This may be proved by breaking something of less value than the magnet—the steel of a lady's stays, for example, hardened and magnetized. It acts like the magnet. When broken, each half acts like the whole; and when these parts are again broken, we have still the perfect magnet, possessing, as in the first instance, two poles. Push your breaking to its utmost limit; you will be driven to prolong your vision beyond that limit, and to contemplate this thing that we call magnetic polarity as resident *in the ultimate particles* of the magnet. Each atom is endowed with this polar force.

Like all other forces, this force of magnetism is amenable to mechanical laws; and, knowing the direction and magnitude of the force, we can predict its action. Placing a small magnetic needle near a bar magnet, it takes up a determinate position. That position might be deduced theoretically from the mutual action of the poles. Moving the needle round the magnet, for each point of the surrounding space there is a definite direction of the needle, and no other. A needle of iron will answer as well as the magnetic needle; for the needle of iron is magnetized by the magnet, and acts exactly like a needle independently magnetized.

If we place two or more needles of iron near the magnet, the action becomes more complex, for then the iron needles are not only acted on by the magnet, but they act upon each other. And if we pass to smaller masses of iron—to iron filings, for example—we find that they act substantially as the needles, arranging themselves in definite forms, in obedience to the magnetic action.

Placing a sheet of paper or glass over this bar magnet and showering iron filings upon the paper, I notice a tendency of the filings to arrange themselves in determinate lines. They cannot freely follow this tendency, for they are hampered by the friction against

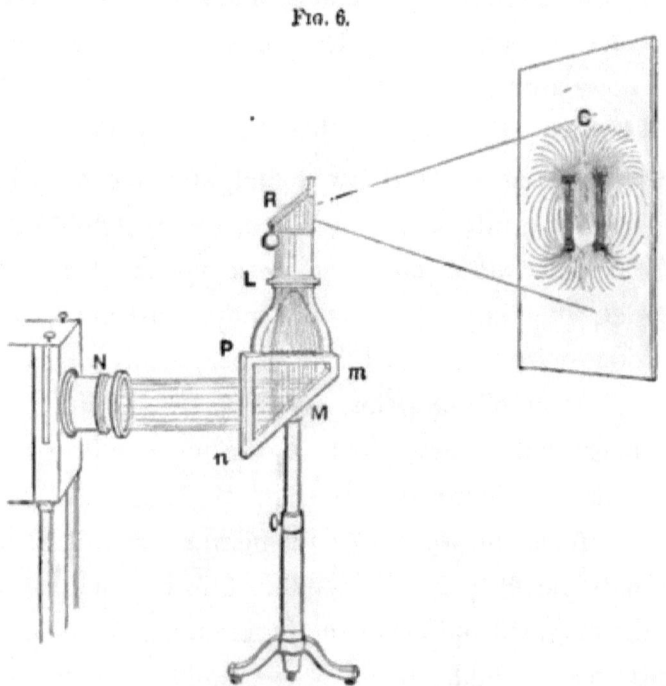

FIG. 6.

N is the nozzle of the lamp; M a plane mirror, reflecting the beam upwards. At P, the magnets and iron filings are placed; L is a lens which forms an image of the magnets and filings; and R is a totally-reflecting prism which casts the image, G, upon the screen.

the paper. They are helped by tapping the paper: each tap releases them for a moment, and enables them to follow their bias. But this is an experiment which can only be seen by myself. To enable you to see it, I take a pair of small magnets and by a simple optical arrangement throw the images of the magnets upon the screen. Scattering iron filings over the glass plate to which the small magnets are attached, and tapping the plate, you see the arrangement of the iron filings in those magnetic curves which have been so long familiar to scientific men.[1]

The aspect of these curves so fascinated Faraday that the greater portion of his intellectual life was devoted to pondering over them. He invested the space through which they run with a kind of materiality; and the probability is, that the progress of science by connecting the phenomena of magnetism with the luminiferous ether, will prove these "lines of force," as Faraday loved to call the magnetic curves, to represent a condition of this mysterious substratum of all radiant action.

But it is not with the magnetic curves, as such, that I now wish to occupy your attention; it is their relationship to theoretic conceptions that we have now to consider. By the action of the bar magnet upon the needle we obtain a notion of a polar force; by the breaking of the strip of magnetized steel, we attain the notion that polarity can attach itself to the ultimate

---

[1] Very beautiful specimens of these curves have been recently obtained, and fixed, by Prof. Mayer, of Hoboken.

particles of matter. The experiment with the iron filings introduces a new idea into the mind; the idea, namely, of *structural arrangement*. Every pair of filings possesses four poles, two of which are attractive and two repulsive. The attractive poles approach, the repulsive poles retreat; the consequence being a certain definite arrangement of the particles with reference to each other.

Now, this idea of structure, as produced by polar force, opens a way for the intellect into an entirely new region, and the reason you are asked to accompany me into this region is, that our next inquiry relates to the action of crystals upon light. Before I speak of this action, I wish you to realize the process of crystalline architecture. Look then into a granite quarry, and spend a few minutes in examining the rock. It is not of perfectly uniform texture. It is rather an agglomeration of pieces, which, on examination, present curiously-defined forms. You have there what mineralogists call quartz, you have felspar, you have mica. In a mineralogical cabinet, where these substances are preserved separately, you will obtain some notion of their forms. You will see there, also, specimens of beryl, topaz, emerald, tourmaline, heavy spar, fluor-spar, Iceland spar—possibly a full-formed diamond, as it quitted the hand of Nature, not yet having got into the hands of the lapidary. These crystals, you will observe, are put together according to law; they are not chance productions; and, if you care to examine them more minutely, you will find their architecture capable of being to some extent revealed. They split in certain directions before a knife-edge, exposing smooth and shining surfaces, which are called planes of cleavage; and by following these planes you sometimes reach an internal form, disguised beneath the external form of the crystal. Ponder these beautiful edifices of a hidden builder. You cannot help asking yourself how they were built; and familiar as you now are with the notion of a polar force, and the ability of that force to produce structural arrangement, your inevitable answer will be, that those crystals are built by the play of polar forces with which their ultimate molecules are endowed. In virtue of these forces, atom lays itself to atom in a perfectly definite way, the final visible form of the crystal depending upon this play of its molecules.

Everywhere in Nature we observe this tendency to run into definite forms, and nothing is easier than to give scope to this tendency by artificial arrangements. Dissolve nitre in water, and allow the water slowly to evaporate; the nitre remains, and the solution soon becomes so concentrated that the liquid form can no longer be preserved. The nitre-molecules approach each other, and come at length within the range of their polar forces. They arrange themselves in obedience to these forces, a minute crystal of nitre being at first produced. On this crystal the molecules continue to deposit themselves from the surrounding liquid. The crystal grows, and finally we have large prisms of nitre, each of a perfectly definite shape. Alum crystallizes with the utmost ease in this fashion. The resultant crystal is, however, different in shape from that of nitre, because the poles of the molecules are differently disposed; and, if they be only *nursed* with proper care, crystals of these substances may be caused to grow to a great size.

The condition of perfect crystallization is, that the crystallizing force shall act with deliberation. There should be no hurry in its operations; but every molecule ought to be permitted, without disturbance from its neighbors, to exercise its own molecular rights. If the crystallization be too sudden, the regularity disappears. Water may be saturated with sulphate of soda, dissolved when the water is hot, and afterward permitted to cool. When cold the solution is supersaturated; that is to say, more solid matter is contained in it than corresponds to its temperature. Still the molecules show no sign of building themselves together. This is a very remarkable, though a very common fact. The molecules in the centre of the liquid are so hampered by the action of their neighbors that freedom to follow their own tendencies is denied to them. Fix your mind's eye upon a molecule within the mass. It wishes to unite with its neighbor to the right, but it wishes equally to unite with its neighbor to the left; the one tendency neutralizes the other, and it unites with neither. We have here, in fact, translated into molecular action, the well-known suspension of animal volition produced by two equally inviting bundles of hay. But, if a crystal of sulphate of soda be dropped into the solution, the molecular indecision ceases. On the crystal the adjacent molecules will immediately precipitate themselves; on these again others will be precipitated, and this act of precipitation will continue from the top of the flask to the bottom, until the solution has, as far as possible, assumed the solid form. The crystals here formed are small, and confusedly arranged. The process has been too hasty to admit of the pure and orderly action of the crystallizing force. It typifies the state of a nation in which natural and healthy change is resisted, until society becomes, as it were, supersaturated with the desire for change, the change being then effected through confusion and revolution, which a wise foresight might have avoided.

Let me illustrate the action of crystallizing force by two examples of it: Nitre might be employed, but another well-known substance enables me to make the experiment in a better form. The substance is common sal-ammoniac, or chloride of ammonium, dissolved in water. Cleansing perfectly a glass plate, the solution of the chloride is poured over the glass, to which, when the plate is set on edge, a thin film of the liquid adheres. Warming the glass slightly, evaporation is promoted; the plate is then placed in a solar microscope, and an image of the film is thrown upon a white screen. The warmth of the illuminating beam adds itself to that already imparted to the glass plate, so that after a moment or two the film can no longer exist in the liquid condition. Molecule then closes with molecule, and you have a most impressive display of crystallizing energy overspreading the whole screen. You may produce something similar if you breathe upon the frost-ferns which overspread your window-panes in winter, and then observe through a lens the subsequent recongelation of the film.

Here the crystallizing force is hampered by the adhesion of the film to the glass; nevertheless, the play of power is strikingly beautiful. Sometimes the crystals start from the edge of the film and run through it from that edge, for, the crystallization being once started, the molecules throw themselves by preference on the crystals already formed. Sometimes the crystals start from definite nuclei in the centre of the film; every small crystalline particle which rests in the film furnishes a starting-point. Throughout the process you notice one feature which is perfectly unalterable, and that is, angular magnitude. The spiculæ branch from the trunk, and from these branches others shoot; but the angles enclosed by the spiculæ are unalterable. In like manner you may find alum-crystals, quartz-crystals, and all other crystals, distorted in shape. They are thus far at the mercy of the accidents of crystallization; but in one particular they assert their superiority over all such accidents—*angular magnitude* is always rigidly preserved.

My second example of the action of crystallizing force is this: By sending a voltaic current through a liquid, you know that we decompose the liquid, and if it contains a metal, we liberate this metal by the electrolysis. This small cell contains a solution of acetate of lead, and this substance is chosen because lead lends itself freely to this crystallizing power. Into the cell dip two very thin platinum wires, and these are connected by other wires with a small voltaic battery. On sending the voltaic current through the solution, the lead will be slowly severed from the atoms with which it is now combined; it will be liberated upon one of the wires, and at the moment of its liberation it will obey the polar forces of its atoms, and produce crystalline forms of exquisite beauty. They are now before you, sprouting like ferns from the wire, appearing indeed like vegetable growths rendered so rapid as to be plainly visible to the naked eye. On reversing the current, these wonderful lead-fronds will dissolve, while from the other wire filaments of lead dart through the liquid. In a moment or two the growth of the lead-trees recommences, but they now cover the other wire. In the process of crystallization, Nature first reveals herself as a builder. Where do her operations stop?

Does she continue by the play of the same forces to form the vegetable, and afterwards the animal? Whatever the answer to these questions may be, trust me that the notions of the coming generations regarding this mysterious thing, which some have called "brute matter," will be very different from those of the generations past.

There is hardly a more beautiful and instructive example of this play of molecular force than that furnished by the case of water. You have seen the exquisite fern-like forms produced by the crystallization of a film of water on a cold window-pane. You have also probably noticed the beautiful rosettes tied together by the crystallizing force during the descent of a snowshower on a very calm day. The slopes and summits of the Alps are loaded in winter with these blossoms of the frost. They vary infinitely in detail of beauty, but the same angular magnitude is preserved throughout. An inflexible power binds spears and spiculæ to the angle of 60 degrees. The common ice of our lakes is also ruled in its deposition by the same angle. You may sometimes see in freezing water small crystals of stellar shapes, each star consisting of six rays, with this angle of 60° between every two of them. This structure may be revealed in ordinary ice. In a sunbeam, or, failing that, in our electric beam, we have an instrument delicate enough to unlock the frozen molecules without disturbing the order of their architecture.

Cutting from clear, sound, regularly-frozen ice a slab parallel to the planes of freezing, and sending a sunbeam through such a slab, it liquefies internally at special points, round each point a six-petalled liquid flower of exquisite beauty being formed. Crowds of such flowers are thus produced.

A moment's further devotion to the crystallization of water will be well repaid; for the sum of qualities which renders this substance fitted to play its part in Nature may well excite wonder and stimulate thought. Like almost all other substances, water is expanded by heat and contracted by cold. Let this expansion and contraction be first illustrated:

A small flask is filled with colored water, and stopped with a cork. Through the cork passes a glass tube water-tight, the liquid standing at a certain height (t', Fig. 7) in the tube. The flask and its tube resemble the bulb and stem of a thermometer. Applying the heat of a spirit-lamp, the water rises in the tube, and finally trickles over the top (t). Expansion by heat is thus illustrated.

Removing the lamp and piling a freezing mixture in the vessel (B) round the flask, the liquid column falls, thus showing the contraction of the water by the cold. But let the freezing mixture continue to act: the falling of the column continues to a certain point; it then ceases. The top of the column remains stationary for some seconds, and afterwards begins to rise. The contraction has ceased, and *expansion by cold* sets in. Let the expansion continue till the liquid trickles a second

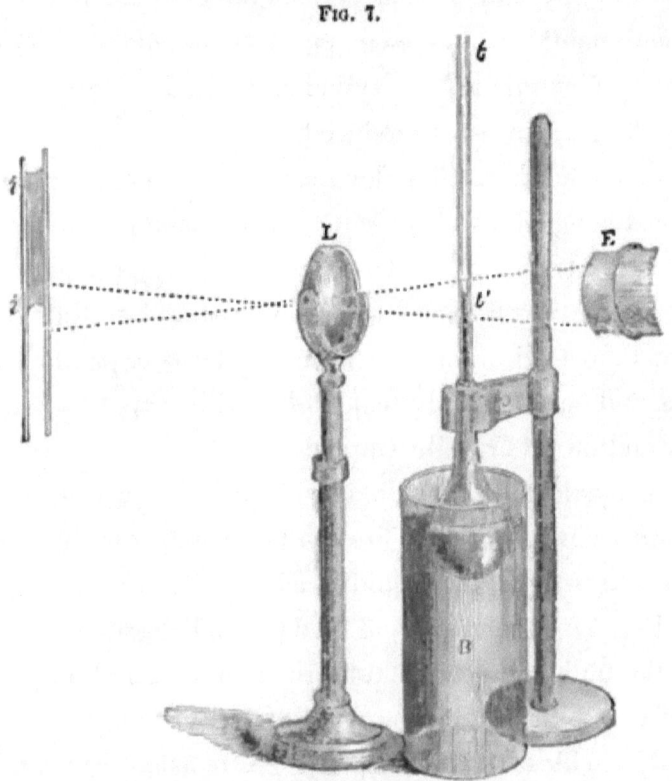

Fig. 7.

Projection of experiment: E is the nozzle of the lamp, L a converging lens, and *i i* the image of the liquid column.

time over the top of the tube. The freezing mixture has here produced to all appearance the same effect as the flame. In the case of water, contraction by cold ceases and expansion by cold sets in at the definite temperature of 39° Fahr. Crystallization has virtually here commenced, the molecules preparing themselves for the subsequent act of solidification which occurs at 32°, and in which the expansion suddenly culminates. In virtue of this expansion, ice, as you know, is lighter than water in the proportion of 8 to 9.[1]

It is my desire, in these lectures, to lead you as closely as possible to the limits hitherto attained by scientific thought, and, in pursuance of this desire, I have now to invite your attention to a molecular problem of great interest, but of great complexity. I wish you to obtain such an insight of the molecular world as shall give the intellect satisfaction when reflecting on the deportment of water before and during the act of crystallization. Consider, then, the ideal case of a number of magnets deprived of weight, but retaining their polar forces. If we had a liquid of the specific gravity of steel, we might, by making the magnets float in it, realize this state of things, for in such a liquid the magnets would neither sink nor swim. Now,

---

[1] In a little volume entitled "Forms of Water," I have mentioned that cold iron floats upon molten iron. In company with my friend Sir William Armstrong, I had repeated opportunities of witnessing this fact in his works at Elswick, in 1863. Faraday, I remember, spoke to me subsequently of the completeness of iron castings as probably due to the swelling of the metal on solidification. Beyond this, I have given the subject no special attention; and I know that many intelligent iron-founders doubt the fact of expansion. It is quite possible that the solid floats because it is not *wetted* by the molten iron, its volume being virtually augmented by capillary repulsion. Certain flies walk freely upon water in virtue of an action of this kind. With bismuth, however, it is easy to burst iron bottles by the force of solidification.

the principle of gravitation is that every particle of matter attracts every other particle with a force varying as the inverse square of the distance. In virtue of the attraction of gravity, then, the magnets, if perfectly free to move, would slowly approach each other.

But besides the unpolar force of gravity, which belongs to matter in general, the magnets are endowed with the polar force of magnetism. For a time, however, the polar forces do not sensibly come into play. In this condition the magnets resemble our water molecules at the temperature say of 50°. But the magnets come at length sufficiently near each other to enable their poles to interact. From this point the action ceases to be a general attraction of the masses. An attraction of special points of the masses and a repulsion of other points now come into play; and it is easy to see that the rearrangement of the magnets consequent upon the introduction of these new forces may be such as to require a greater amount of room. This, I take it, is the case with our water-molecules. Like the magnets, they approach each other *as wholes*, until the temperature 39° is reached. Previous to this temperature, doubtless, the polar forces had begun to act, and at this temperature their action exactly balances the contraction due to cold. At lower temperatures the polar forces predominate. But they carry on a gradual struggle with the force of contraction until the freezing temperature is attained. Here the polar forces suddenly and finally gain the victory. The molecules close up and form solid crystals, a considerable augmentation of volume being the immediate consequence.

We can still further satisfy the intellect by showing that these conceptions can be realized by a model. The molecule of water is composed of two atoms of hydro-

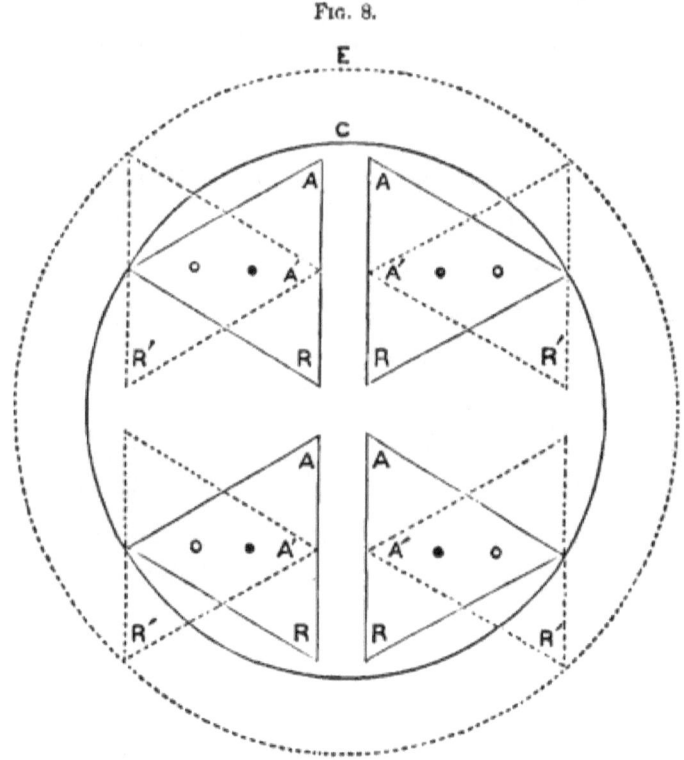

Fig. 8.

gen, united to one of oxygen. We may assume the *molecule* built up of these atoms to be pyramidal. Suppose the triangles in Fig. (8) to be drawn touching the sides of the molecule, and the disposition of the polar forces to be that indicated by the letters; the points marked A being attractive, and those marked R repellent. In virtue of the *general* attraction of the molecules, let them be drawn towards the positions marked by the *full* lines, and then suppose the polar attractions and repulsions to act. A will turn towards A, and R will retreat from R. The molecules will be caused to rotate, their final position being that shown by the *dotted* lines. But the circle surrounding the latter is larger than that surrounding the full lines, which shows that the molecules in their new positions require more room. In this way we obtain an image of the molecular mechanism active in the case of water. The demand for more room is made with an energy sufficient to overcome all ordinary resistances. Your lead pipes yield readily to this power; but iron does the same, and bomb-shells, as you know, can be burst by the freezing of water. Thick iron bottles filled with water and placed in a freezing mixture are shivered into fragments by the resistless vigor of molecular force.

We have now to exhibit the bearings of crystallization upon optical phenomena. According to the undulatory theory, the velocity of light in water and glass is less than in air. Consider, then, a small portion of a wave issuing from a point of light so distant that the portion may be regarded as practically straight. Moving vertically downwards, and impinging on an horizontal surface of glass, the wave would go through the glass without change of direction. But, as the velocity in glass is less than the velocity in air, the wave would be retarded on passing into the denser medium.

But suppose the wave, before reaching the glass, to be *oblique* to the surface; that end of the wave which first reaches the glass will be the first retarded, the other portions as they enter the glass being retarded in succession. This retardation of the one end of the wave causes it to swing round and change its front, so that when the wave has fully entered the glass its course is oblique to its original direction. According to the undulatory theory, light is thus *refracted*.

The two elements of rapidity of propagation, both of sound and light, in any substance whatever, are *elasticity* and *density*, and the enormous velocity of light is attainable because the ether is at the same time of infinitesimal density and of enormous elasticity. It surrounds the atoms of all bodies, but seems to be so acted upon by them that its density is increased without a proportionate increase of elasticity; this would account for the diminished velocity of light in refracting bodies. In virtue of the crystalline architecture that we have been considering, the ether in many crystals possesses different densities in different directions;

and the consequence is, that some of these media transmit light with two different velocities. Now, refraction depends wholly upon the change of velocity on entering the refracting medium; and is greatest where the change of velocity is greatest. Hence, as, in many crystals, we have two different velocities, we have also two different refractions, a beam of light being divided by such crystals into two. This effect is called *double refraction*.

In water, for example, there is nothing in the grouping of the molecules to interfere with the perfect homogeneity of the ether; but, when water crystallizes to ice, the case is different. In a plate of ice the elasticity of the ether in a direction perpendicular to the surface of freezing is different from what it is parallel to the surface of freezing; ice is, therefore, a double refracting substance. Double refraction is displayed in a particularly impressive manner by Iceland spar, which is crystallized carbonate of lime. The difference of ethereal density in two directions in this crystal is very great, the separation of the beam into the two halves being, therefore, particularly striking.

Before you is now projected an image of our carbon-points. Introducing the spar, the beam which builds the image is permitted to pass through it; instantly you have the single image divided into two. Projecting an image of the aperture through which the light issues from the electric lamp, and introducing the spar, two luminous disks, instead of one, appear immediately upon the screen. (See Fig. 9.)

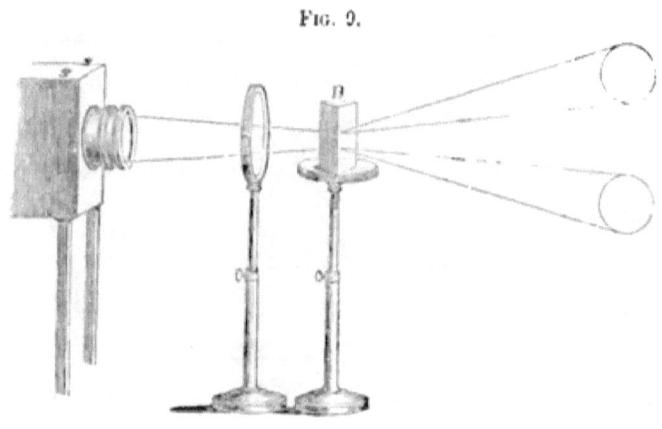

Fig. 9.

The two beams into which the spar divides the single incident-beam do not behave alike. One of them obeys the ordinary law of refraction discovered by Snell, and this is called *the ordinary ray*. The other does not obey the ordinary law. Its index of refraction, for example, is not constant, nor do the incident and refracted rays always lie in the same plane. It is, therefore, called *the extraordinary ray*. Pour water and bisulphide of carbon into two cups of the same depth; looked at through the liquid, the cup that contains the more strongly-refracting liquid will appear shallower than the other. Place a piece of Iceland spar over a dot of ink; two dots are seen, but one appears nearer than the other. The nearest dot belongs to the most strongly-refracted ray, which in this case is the ordinary ray. Turn the spar round, and the extraordinary image of the spot rotates round the ordinary one.

The double refraction of Iceland spar was first treated in a work published by Erasmus Bartholinus, in 1669. The celebrated Huyghens sought to account for the phenomenon on the principles of the wave theory, and he succeeded in doing so. He made highly important observations on the distinctive character of the two beams transmitted by the spar. Newton, reflecting on the observations of Huyghens, came to the conclusion that each of the beams had two sides; and from the analogy of this *two-sidedness* with the *two-endedness* of a magnet, wherein consists its polarity, the two beams came subsequently to be described as *polarized*.

We shall study this subject of the *polarization of light* with great ease and profit by means of a crystal of tourmaline. But let us start with a clear conception of an ordinary beam of light. It has been already explained that the vibrations of the individual ether-particles are executed *across* the line of propagation. In the case of ordinary light we are to figure the ether particles as vibrating in all directions, or azimuths, as it is sometimes expressed, across this line.

Now, in a plate of tourmaline cut parallel to the axis of the crystal, the beam of incident light is divided into two, the one vibrating parallel to the axis of the crystal, the other at right angles to the axis. The grouping of the molecules, and of the ether associated with the molecules, reduces all the vibrations incident upon the crystal to these two directions. One of these beams, namely that one whose vibrations are perpendicular to the axis, is quenched with exceeding rapidity by the tourmaline, so that, after having passed through a very small thickness of the crystal, the light emerges with all its vibrations reduced to a single plane. In this condition it is what we call a beam of *plane polarized light.*

A moment's reflection will show, if what has been stated be correct, that, on placing a second plate of tourmaline with its axis parallel to the first, the light will pass through both; but that, if the axes be crossed,

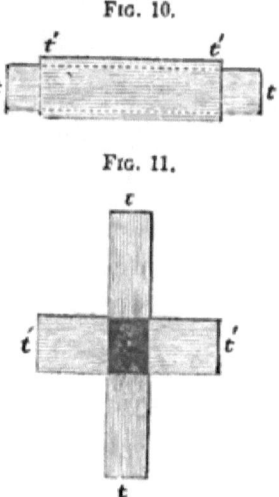

Fig. 10.

Fig. 11.

the light that passes through the one plate will be quenched by the other, a total interception of the light being the consequence. The image of a plate of tourmaline, tt (Fig. 10), is now before you. I place parallel to it another plate, $t't'$: the green of the crystal is a little deepened, nothing more. By means of an endless screw, I now turn one of the crystals gradually round; as long as the two plates are oblique to each other, a certain portion of light gets through; but, when they are at right angles to each other, the space common to both is a space of darkness, as shown in Fig. 11.

Let us return to a single plate; and let me say that it is on the green light transmitted by the tourmaline that you are to fix your attention. We have now to illustrate the two-sidedness of that green light. The light surrounding the green image being ordinary light, is reflected by a plane glass mirror in all directions; the green light, on the contrary, is not so reflected. The image of the tourmaline is now horizontal; reflected upwards, it is still green; reflected sideways, the image is reduced to blackness, because of the incompetency of the green light to be reflected in this direction. Making the plate of tourmaline vertical and reflecting it as before, in the upper image the light is quenched; in the side image you have now the green. Picture the thing clearly. In the one case the mirror receives the impact of the *edges* of the waves, and the green light is quenched. In the other case the *sides* of the waves strike the mirror, and the green light is reflected. To render the extinction complete, the light must be received upon the mirror at a special angle. What this angle is we shall learn presently.

The quality of two-sidedness conferred upon light by crystals may also be conferred upon it by ordinary reflection. Malus made this discovery in 1808, while looking through Iceland spar at the light of the sun reflected from the windows of the Luxembourg palace in Paris. I receive upon a plate of window-glass the beam from our lamp; a great portion of the light reflected from the glass is polarized; the vibrations of this reflected beam are executed, for the most part, parallel to the surface of the glass, and, if the glass be held so that the beam shall make an angle of 58° with the perpendicular to the glass, the *whole* of the reflected beam is polarized. It was at this angle that the image of the tourmaline was completely quenched in our former experiments. It is called *the polarizing angle*.

And now let us try to make substantially the experiment of Malus. I receive the beam from the lamp upon this plate of glass and reflect it through the spar. Instead of two images, you see but one. So that the light, when polarized, as it now is, can only get through the spar in one direction, and consequently produce but one image. Why is this? In the Iceland spar, as in the tourmaline, all the vibrations of the ordinary light are reduced to two planes at right angles to each other; but, unlike the tourmaline, both beams are transmitted with equal facility by the spar. The two beams, in short, emergent from the spar, are polarized, their directions of vibration being at right angles to each other. When, therefore, the light was polarized by reflection, the direction of vibration in the spar which corresponded to the direction of vibration of the polarized beam transmitted it, and that direction only. But one image, therefore, was possible under the conditions.

And now you have it in your power to check many of my statements, and you will observe that such logic as connects our experiments is simply a transcript of the logic of Nature. On the screen before you are the two disks of light produced by the double refraction of the spar. They are, as you know, two images of the aperture through which the light issues from the camera. Placing the tourmaline in front of the aperture, two images of the crystal will be obtained; but now let us reason out what is to be expected from this experiment. The light emergent from the tourmaline is polarized. Placing the crystal with its axis horizontal, the vibrations of the transmitted light will be horizontal. Now the spar, as already stated, has two perpendicular directions of vibration, one of which at the present moment is vertical, the other horizontal. What are we to conclude? That the green light will be transmitted along the latter, which is parallel to the tourmaline, and not along the former, which is perpendicular to it. Hence we may infer that one image of

EXPERIMENTS WITH TOURMALINES.

the tourmaline will show the ordinary green light of the crystal, while the other image will be black. Let us test our reasoning by experiment: it is verified to the letter. (Fig. 12.)

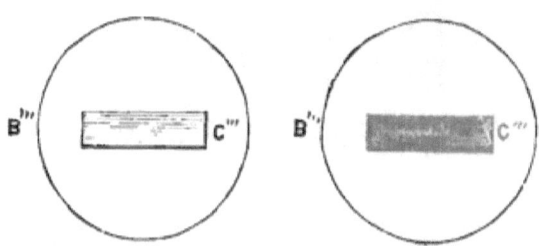

FIG. 12.

Let us push our test still further. By means of an endless screw, the crystal can be turned ninety degrees round. The black image, as I turn, becomes gradually brighter, and the bright one gradually darker; at an angle of forty-five degrees both images are equally

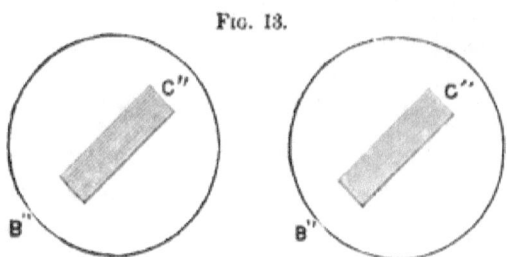

FIG. 13.

bright (Fig. 13); while, when ninety degrees have been obtained, the axis of the crystal being then vertical, the bright and black images have changed places. (Fig. 14.)

Given two beams transmitted through Iceland spar,

it is perfectly manifest that we have it in our power to determine instantly, by means of a plate of tourmaline,

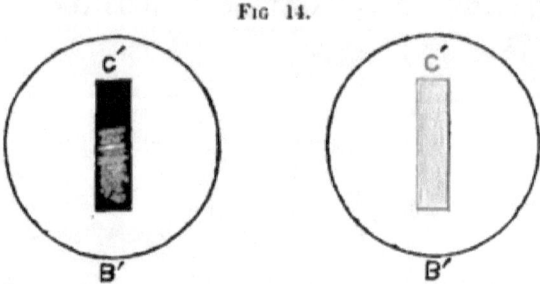

FIG 14.

the directions in which the ether-particles vibrate in the two beams. I might place the double-refracting spar in any position whatever. A minute's trial with the tourmaline would enable you to determine the position which yields a black and a bright image, and from these you would at once infer the directions of vibration

Further, the two beams from the spar being thus polarized, if they be suitably received upon a plate of glass at the polarizing angle, one of them will be reflected, the other not. This is the conclusion of reason from our previous knowledge; but you observe that reason is justified by experiment. (Figs. 15 and 16.)

I have said that the whole of the beam reflected from glass at the polarizing angle is polarized; a word must now be added regarding the larger portion of the light *transmitted* by the glass. The transmitted beam contains a quantity of polarized light equal to that of the reflected beam; but this quantity is only a fraction of the whole transmitted light. By taking two plates of glass instead of one, we augment

FIG. 15.

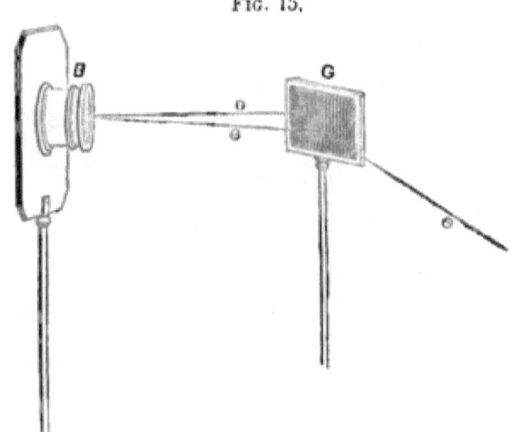

(B is the birefracting spar, dividing the incident light into the two beams o and e. G is the mirror.) The beam is here reflected *laterally*. When the reflection is *upwards*, the other beam is reflected as shown in Fig. 16.

FIG. 16.

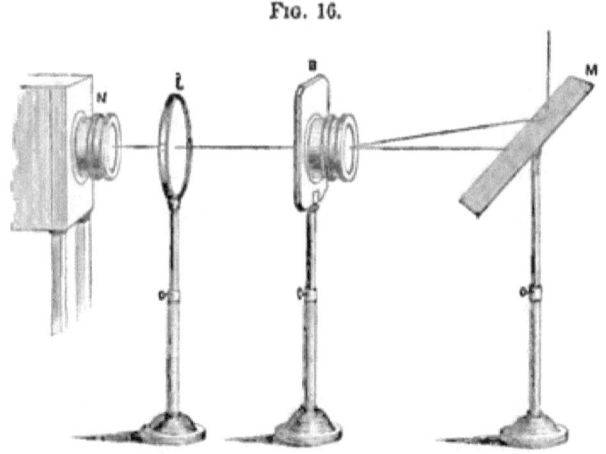

the quantity of the transmitted polarized light; and, by taking *a bundle* of plates, we so increase the quantity as to render the transmitted beam, for all practical purposes, *perfectly* polarized. Indeed, bundles of glass plates are often employed as a means of furnishing polarized light.

One word more. When the tourmalines are crossed, the space where they cross each other is black. But we have seen that the least obliquity on the part of the crystals permits light to get through both. Now suppose, when the two plates are crossed, that we interpose a third plate of tourmaline between them, with its axis oblique to both. A portion of the light transmitted by the first plate will get through this intermediate one. But, after it has got through, *its plane of vibration is changed:* it is no longer perpendicular to the axis of the crystal in front. Hence it will get through that crystal. Thus, by reasoning, we infer that the interposition of a third plate of tourmaline will in part abolish the darkness produced by the perpendicular crossing of the other two plates. I have not a third plate of tourmaline; but the talc or mica which you employ in your stoves is a more convenient substance, which acts in the same way. Between the crossed tourmalines, I introduce a film of this crystal. You see the edge of the film slowly descending, and, as it descends between the tourmalines, light takes the place of darkness. The darkness, in fact, seems scraped away as if it were something material. This effect has been called—and improperly called—*depolarization*.

LECTURE IV.

Chromatic Phenomena produced by Crystals on Polarized Light: The Nicol Prism: Polarizer and Analyzer: Action of thick and thin Plates of Selenite: Colors dependent on Thickness: Resolution of Polarized Beam into two others by the Selenite: One of them more retarded than the other: Recompounding of the two Systems of Waves by the Analyzer: Interference thus rendered possible: Consequent Production of Colors: Action of Bodies Mechanically strained or pressed: Action of Sonorous Vibrations: Action of Glass strained or pressed by Heat: Circular Polarization: Chromatic Phenomena produced by Quartz: The Magnetization of Light: Rings surrounding the Axes of Crystals: Biaxal and Uniaxal Crystals: Grasp of the Undulatory Theory.

We now stand upon the threshold of a new and splendid optical domain. We have to examine, this evening, the chromatic phenomena produced by the action of crystals, and double-refracting bodies generally, upon polarized light. For a long time investigators were compelled to employ plates of tourmaline for this purpose, and the progress they made with so defective a means of inquiry is astonishing. But these men had their hearts in their work, and were on this account enabled to extract great results from small instrumental appliances. But we have better apparatus now. You have seen the two beams emergent from Iceland spar, and have proved them to be polarized. If we could abolish one of these beams, we might employ the other for experiments on polarized light.

These beams, as you know, are refracted differently, and from this we are able to infer that under some circumstances the one may be totally reflected, and the other not. An optician, named Nicol, cut a crystal of Iceland spar in two in a certain direction. He polished the severed surfaces, and reunited them by Canada balsam, the surface of union being so inclined to the beam traversing the spar that the ordinary ray, which is the most highly refracted, was totally reflected by the balsam, while the extraordinary ray was permitted to pass on. The invention of the Nicol prism was a great step in practical optics, and quite recently such prisms have been constructed of a size which enables audiences like the present to witness the chromatic phenomena of polarized light to a degree altogether unattainable a short time ago. The two prisms here before you belong to my excellent friend Mr. William Spottiswoode, and they were manufactured by Mr. Ladd. I have with me another pair of very noble prisms, still larger than these, manufactured for me by Mr. Browning, who has gained so high and well-merited a reputation in the construction of spectroscopes.

These two Nicol prisms play the same part as the crystals of tourmaline. Placed with their directions of vibration parallel, the light passes through both. When these directions are crossed, the light is quenched.

Introducing a film of mica between the prisms, the light is in part restored. But notice, when the film of mica is *thin*, you have sometimes not only light, but *colored* light. Our work for some time to come will be the examination of these colors. With this view, I will take a representative crystal, one easily dealt with; the crystal gypsum, or selenite, which is crystallized sulphate of lime. Between the crossed Nicols I place a thick plate of this crystal; like the mica, it restores the light, but it produces no color. With my penknife I take a thin splinter from this crystal and place it between the prisms; its image on the screen glows with the richest colors. Turning the prism in front, these colors gradually fade, disappear, but, by continuing the rotation until the vibrating sections of the prisms are parallel, vivid colors again appear, but these colors are complementary to the former ones.

Some patches of the splinter appear of one color, some of another. These differences are due to the different thicknesses of the film. If the thickness be uniform, the color is uniform. Here, for instance, is a stellar shape, every lozenge of the star being a film of gypsum of uniform thickness. Each lozenge, you observe, shows a brilliant uniform color. It is easy, by shaping our films so as to represent flowers or other objects, to exhibit such objects in colors unattainable by art. Here, for example, is a specimen of heart's-ease, the colors of which you might safely defy the artist to reproduce. By turning the front Nicol 90 degrees round, we pass through a colorless phase to a series of colors complementary to the former ones. Here, for example, is a rose-tree with a red flower and green leaves; turning the prism 90 degrees round, we obtain a green flower and red leaves. All these wonderful chromatic effects have definite mechanical causes in the motions of the ether. The principle of interference, duly applied and interpreted, explains them all.

By this time you have learned that the word "light" may be used in two different senses: it may mean the impression made upon consciousness, or it may mean the physical agent which makes the impression. It is with the agent that we have to occupy ourselves at present. That agent is the motion of a substance which fills all space, and surrounds the atoms and molecules of bodies. To this interstellar and interatomic medium definite mechanical properties are ascribed, and we deal with it as a body possessed of these properties. In mechanics we have the composition and resolution of forces, and of motions, extending to the composition and resolution of *vibrations*. We treat the luminiferous ether on mechanical principles, and, from the composition, resolution, and interference of its vibrations, we deduce all the phenomena displayed by crystals in polarized light.

Let us take, as an example, the crystal of tourmaline, with which we are now so familiar. Let a vibration cross this crystal oblique to its axis; we have seen by experiment that a portion of the light will pass through. How much, we determine in this way: Draw a straight line representing the intensity of the vibration before it reaches the tourmaline, and from the two ends of this line draw two perpendiculars to the axis of the crystal; the distance between the feet of these two perpendiculars will represent the intensity of the transmitted vibration.

Follow me now while I endeavor to make clear to you what occurs when a film of gypsum is placed between the Nicol prisms. But, at the outset, let us establish still further the analogy between the action of the prisms and that of two plates of tourmaline. The plates are now crossed, and you see that, by turning the film round, it may be placed in a position where it has no power to abolish the darkness. Why is this? The answer is, that in the gypsum there are two directions, at right angles to each other, which the waves of light are constrained to follow, and that now one of these directions is parallel to one of the axes of the tourmaline, and the other parallel to the other axis. When this is the case, the film exercises no sensible action upon the light. But now I turn the film so as to render its direction of vibration *oblique* to the axes; then you see it has the power, demonstrated in the last lecture, of restoring the light.

Let us now mount our Nicol prisms, and cross

them as we crossed the tourmalines. Introducing our film of gypsum between them, you notice that in one particular position the film has no power whatever over the field of view. But, when the film is turned a little way round, the light passes. We have now to understand the mechanism by which this is effected.

FIG. 17

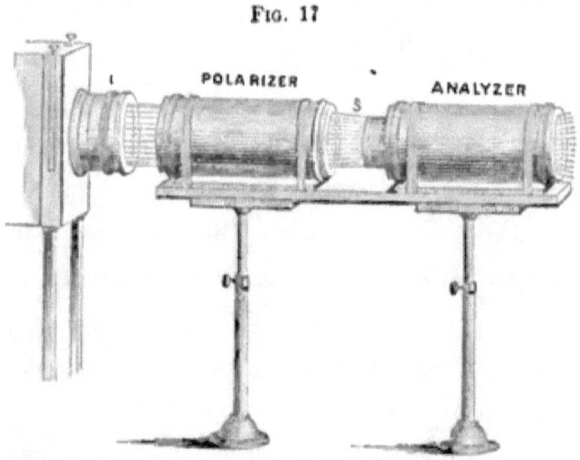

Firstly, then, we have this first prism which receives the light emergent from the electric lamp, and which is called the *polarizer*. Then we have the plate of gypsum, placed at S (Fig. 17), and then the prism in front, which is called the *analyzer*. On its emergence from the first prism, the light is polarized; and, in the particular case now before us, its vibrations are executed in an horizontal plane. The two directions of vibration of the gypsum, placed at S, are now oblique to the horizon. Draw a rectangular cross upon paper to represent the two directions of vibration within the gypsum. Draw an oblique line to represent the intensity of the vibration when it reaches the gypsum. Let fall from the two ends of this line two perpendiculars on each of the arms of the cross; then the distances between the feet of these perpendiculars represent the intensities of two rectangular vibrations *which are the equivalents of the first single vibration.* Thus the polarized ray, when it enters the gypsum, is resolved into two others, vibrating at right angles to each other.

Now, in one of those directions of vibration the ether is more sluggish than in the other; and, as a consequence, the waves that follow this direction are more retarded than the others. The waves of both systems, in fact, are *shortened* when they enter the gypsum, but the one system is *more* shortened than the other. You can readily imagine that in this way the one system of waves may get half a wave-length, or indeed any number of half wave-lengths, in advance of the other. The possibility of interference here flashes upon the mind. A little consideration, however, renders it evident that, as long as the vibrations are executed at right angles to each other, they cannot quench each other, no matter what the retardation may be. This brings us at once to the part played by the analyzer. Its sole function is to recompound the two vibrations emergent from the gypsum. It reduces them to a single plane, where, if one of them be retarded by the proper amount, extinction can occur. But here, as in the case of thin films, the different lengths of the waves of light come into play. Red will require a greater thickness to produce the retardation necessary for extinction than blue; consequently, when the longer waves have been withdrawn by interference, the shorter ones remain and confer their colors on the film of gypsum. Conversely, when the shorter waves have been withdrawn, the thickness is such that the longer waves remain. An elementary consideration suffices to show that, when the directions of vibration of prisms and gypsum enclose an angle of forty-five degrees, the colors are at their maximum brilliancy. When the film is turned from this direction, the colors gradually fade, until, at the point where the directions are parallel, they disappear altogether.

A knowledge of these phenomena is best obtained by means of a model of wood or pasteboard representing the plate of gypsum, its planes of vibration, and also those of the polarizer and analyzer. On these planes the waves may be drawn, showing the resolution of the first polarized ray into two others, and then the reduction of the two vibrations to a common plane. Following out rigidly the interaction of the two systems of waves, we are taught by such a model that all the phenomena of color, obtained when the planes of vibration of the two Nicols are parallel, are displaced by the *complementary* phenomena when the Nicols are perpendicular to each other.

In considering the next point, for the sake of simplicity, we will operate with monochromatic light—with red light, for example. Supposing that a certain thickness of the gypsum produces a retardation of half a wave-length, twice this thickness will produce a retardation of two half wave-lengths; three times this thickness a retardation of three half wave-lengths, and so on. Now, when the Nicols are parallel, the retardation of half a wave-length, or of any *odd* number of half wave-lengths, produces extinction; at all thicknesses, on the other hand, which correspond to a retardation of an *even* number of half wave-lengths, the two beams support each other, when they are brought to a common plane by the analyzer. Supposing, then, that we take a plate of a wedge-form, which grows gradually thicker from edge to back, we ought to expect in red light a series of recurrent bands of light and darkness; the dark bands occurring at thicknesses which produce retardations of one, three, five, etc., half wave-lengths, while the light bands occur between the dark ones. Experiment proves the wedge-shaped crystal to show these bands; but they are far better shown by this circular film, which is so worked as to be thinnest at the centre, gradually increasing in thickness from the centre outwards. These splendid rings of light and darkness are thus produced.

When, instead of employing red light, we employ blue, the rings are also seen; but, as they occur at thinner portions of the film, they are smaller than the rings obtained with the red light. The consequence of employing *white* light may be now inferred: inasmuch as the red and the blue fall in different places, we have *iris-colored* rings produced by the white light.

Some of the chromatic effects of irregular crystallization are beautiful in the extreme. Could I introduce between our Nicols a pane of glass covered by those frost-ferns which the cold weather renders now so frequent, rich colors would be the result. The beautiful effects of irregular crystallization on glass plates, now presented to you, illustrate what you might expect from the frosted window-pane. And not only do crystalline bodies act thus upon light, but almost all bodies that possess a definite structure do the same. As a general rule, organic bodies act in this way; for their architecture implies an arrangement of the ether which involves double refraction. A film of horn, or the section of a shell, for example, yields very beautiful colors in polarized light. In a tree, the ether certainly possesses different degrees of elasticity along and across the fibre; and, were wood transparent, this peculiarity of molecular structure would infallibly reveal itself by chromatic phenomena like those that you have seen. But not only do bodies built permanently by Nature behave in this way, but it is possible, as shown by Brewster, to confer, by strain or by pressure, a temporary double-refracting structure upon non-crystalline bodies, such as common glass.

When I place this bar of wood across my knee and seek to break it, what is the mechanical condition of the bar? It bends, and its convex surface is *strained* longitudinally; its concave surface, that next my knee, is longitudinally *pressed*. Both in the strained portion and in the pressed portion the ether is thrown into a condition which would render the wood, were it transparent, double-refracting. Let us repeat the experiment with a bar of glass. Between the crossed Nicols I introduce such a bar. By the dim residue of light lingering upon the screen, you see the image of the glass, but it has no effect upon the light. I simply bend the glass bar with my finger and thumb, keeping its length oblique to the directions of vibration in the Nicols. Instantly light flashes out upon the screen. The two sides of the bar are illuminated, the edges most, for here the strain and pressure are greatest. In passing from strain to pressure, we cross a portion of the glass where neither is exerted. This is the so-called neutral axis of the bar of glass, and along it you see a dark band, indicating that the glass along this axis exercises no action upon the light. By employing the force of a press, instead of the force of my finger and thumb, the brilliancy of the light is greatly augmented.

Again, I have here a square of glass which can be

inserted into a press of another kind. Introducing the square between the prisms, its neutrality is declared; but it can hardly be held sufficiently loosely to prevent its action from manifesting itself. Already, though the pressure is infinitesimal, you see spots of light at the points where the press is in contact with the glass. I now turn this screw. Instantly the image of the square of glass flashes out upon the screen. You see luminous spaces separated from each other by dark bands. Every pair of adjacent luminous spaces is in opposite mechanical conditions. On one side of the dark band we have strain, on the other side pressure; while the dark band marks the neutral axis between both. I now tighten the vice, and you see color; tighten still more, and the colors appear as rich as those presented by crystals. Releasing the vice, the colors suddenly vanish; tightening suddenly, they reappear. From the colors of a soap-bubble Newton was able to infer the thickness of the bubble, thus uniting by the bond of thought apparently incongruous things. From the colors here presented to you, the magnitude of the pressure employed might be inferred. Indeed, the late M. Wertheim, of Paris, invented an instrument for the determination of strains and pressures by the colors of polarized light, which exceeded in accuracy all other instruments of the kind.

You know that bodies are expanded by heat and

contracted by cold. If the heat be applied with perfect uniformity, no local strains or pressures come into play; but, if one portion of a solid be heated and others not, the expansion of the heated portion introduces strains and pressures which reveal themselves under the scrutiny of polarized light. When a square of common window-glass is placed between the Nicols, you see its dim outline, but it exerts no action on the polarized light. Held for a moment over the flame of a spirit-lamp, on reintroducing it between the Nicols, light flashes out upon the screen. Here, as in the case of mechanical action, you have spaces of strain divided by neutral axes from spaces of pressure.

Let us apply the heat more symmetrically. This small square of glass is perforated at the centre, and into the orifice a bit of copper wire is introduced. Placing the square between the prisms, and heating the copper, the heat passes by conduction along the wire to the glass, through which it spreads from the centre outwards. You see a dim cross bounding four luminous quadrants growing up and becoming gradually black by comparison with the adjacent brightness. And as, in the case of pressure, we produced colors, so here also, by the proper application of heat, gorgeous chromatic effects may be produced. And they may be rendered permanent by first heating the glass sufficiently, and then cooling it, so that the chilled mass shall remain in a state of strain and pressure. Two or three examples will illustrate this point. The colors, you observe, are quite as rich as those obtained in the case of crystals.

And now we have to push these considerations to a final illustration. Polarized light may be turned to account in various ways as an analyzer of molecular condition. A strip of glass six feet long, two inches wide, and a quarter of an inch thick, is held at the centre between my finger and thumb. I sweep over one of its halves a wet woollen rag; you hear an acute sound, due to the vibrations of the glass. What is the condition of the glass while the sound is heard? This: its two halves lengthen and shorten in quick succession. Its two ends, therefore, are in a state of quick vibration; but at the centre the pulses from the two ends alternately meet and retreat. Between their opposing actions, the glass at the centre is kept motionless; but, on the other hand, it is alternately strained and compressed. The state of the glass may be illustrated by a row of spots of light, as the propagation of a sonorous pulse was illustrated in a former lecture. By a simple mechanical contrivance the spots are made to vibrate to and fro. The terminal dots have the largest amplitude of vibration, while those at the centre are alternately crowded together and drawn asunder, the centre one not moving at all. The condition of the sounding strip of glass is here correctly represented. In Fig. 18, A B represents the glass rectangle with its centre condensed; while A′ B′ represents the same rectangle with its centre rarefied.

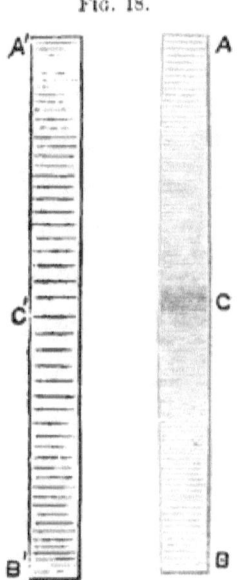

Fig. 18.

If we introduce the glass $s\,s'$ (Fig. 19) between the crossed Nicols, taking care to keep the strip oblique to the direction of vibration of the Nicols, and sweep our wet rubber over the glass, this may be expected to occur: At every moment of compression the light will flash through; at every moment of strain the light will also flash through; and these states of strain and pressure will follow each other so rapidly that we may expect a permanent luminous impression to be made upon the eye. By pure reasoning, therefore, we reach the conclusion that the light will be revived whenever the glass is sounded. That it is so, experiment testifies: at every sweep of the rubber, a fine luminous disk (o) flashes out upon the screen. The experiment may be varied in this way: Placing in front of the polarizer a plate of unannealed glass, you have those beautiful colored rings, intersected by a black cross. Every sweep of the rubber not only abolishes the rings, but introduces complementary ones, the black cross being for the moment supplanted by a white one. This is a modification of an experiment which we owe to Biot. His apparatus, however, confined the observation of it to a single person at a time.

Fig. 19.

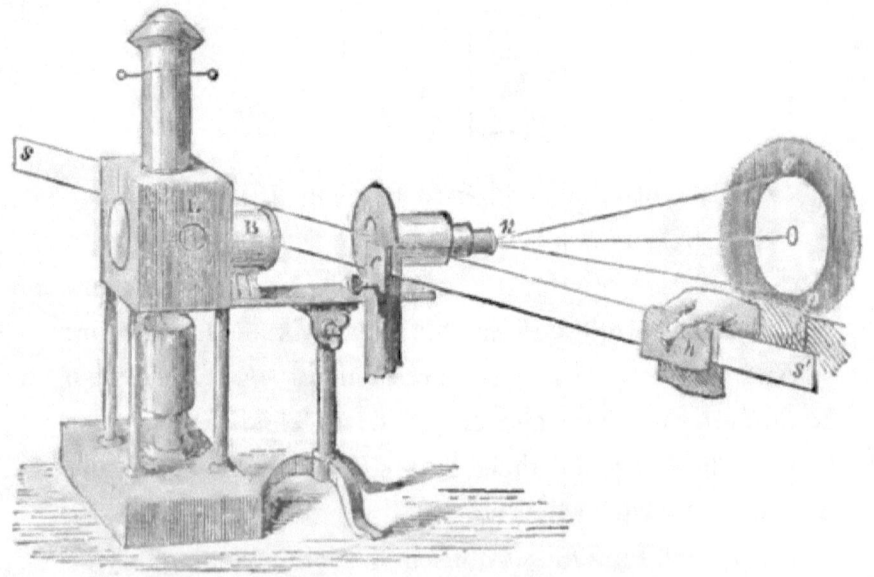

But we have to follow the ether still further. Suspended before you is a pendulum, which, when drawn aside and then liberated, oscillates to and fro. If, when the pendulum is passing the middle point of its excursion, I impart a shock to it tending to drive it at right angles to its present course, what occurs? The two impulses compound themselves to a vibration oblique in direction to the former one, but the pendulum oscillates in *a plane.* But, if the rectangular shock be imparted to the pendulum when it is at the limit of its swing, then the compounding of the two impulses causes the suspended ball to describe not a straight line, but an ellipse; and, if the shock be competent of itself to produce a vibration of the same amplitude as the first one, the ellipse becomes a circle. But why do I dwell upon these things? Simply to make known to you the resemblance of these gross mechanical vibrations to the vibrations of light. I hold in my hand a plate of quartz cut from the crystal perpendicular to its axis. This crystal thus cut possesses the extraordinary power of twisting the plane of vibration of a polarized ray to an extent dependent on the thickness of the crystal. And the more refrangible the light the greater is the amount of twisting, so that, when white light is employed, its constituent colors are thus drawn asunder. Placing the quartz between the polarizer and the analyzer, you see this splendid color, and, turning the analyzer in front, from right to left, the other colors appear in succession. Specimens of quartz have been found which require the analyzer to be turned from left to right, to obtain the same succession of colors. Crystals of the first class are therefore called right-handed, and, of the second class, left-handed crystals.

With profound sagacity, Fresnel, to whose genius we mainly owe the expansion and final triumph of the undulatory theory of light, reproduced mentally the mechanism of these crystals, and showed their action to be due to the circumstance that, in them, the waves of ether so act upon each other as to produce the condition represented by our rotating pendulum. Instead of being plane polarized, the light in rock crystal is *circularly polarized*. Two such rays transmitted along the axis of the crystal, and rotating in opposite directions, when brought to interference by the analyzer, are demonstrably competent to produce the observed phenomena.

I now abandon the analyzer, and put in its place the piece of Iceland spar with which we have already illustrated double refraction. The two images of the carbon-points are now before you. Introducing a plate of quartz between the polarizer and the spar, the two images glow with complementary colors. Employing the image of an aperture instead of that of the carbon-points, we have two complementary colored circles. As the analyzer is caused to rotate, the colors pass through various changes; but they are always complementary to each other. If the one be red, the other will be

MIXTURE OF YELLOW AND BLUE. 119

green; if the one be yellow, the other will be blue. Here we have it in our power to demonstrate afresh a statement made in a former lecture, that, although the mixture of blue and yellow pigments produces green, the mixture of blue and yellow lights produces white. By enlarging our aperture, the two images produced by the spar are caused to approach each other, and finally to overlap. The one is now a vivid yellow, the other a vivid blue, and you notice that where the colors are superposed we have a pure white. (See Fig. 20, where N is the nozzle of the lamp, Q the quartz plate, L a lens, and B the birefracting spar. The two images overlap at O, and produce white by their mixture.)

Fig. 20.

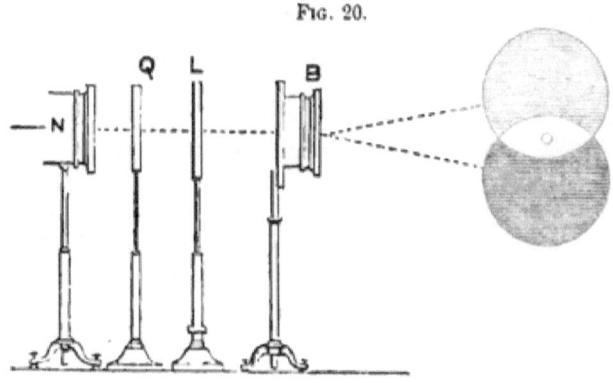

This brings us to a point of our inquiries which, though not capable of brilliant illustration, is nevertheless so likely to affect profoundly the future course of scientific thought that I am unwilling to pass it over without reference. I refer to the experiment which

Faraday, its discoverer, called the *magnetization of light*. The arrangement for this celebrated experiment is now before you. We have first our electric lamp, then a Nicol prism, to polarize the beam emergent from the lamp; then an electro-magnet, then a second Nicol prism, and finally our screen. At the present moment the prisms are crossed, and the screen is dark. I place from pole to pole of the electro-magnet a cylinder of a peculiar kind of glass, first made by Faraday, and called Faraday's heavy glass. Through this glass the beam from the polarizer now passes, being intercepted by the Nicol in front. I now excite the magnet, and instantly light appears upon the screen. On examination, we find that, by the action of the magnet upon the ether contained within the heavy glass, the plane of vibration is caused to rotate, thus enabling the light to get through the analyzer.

The two classes into which quartz-crystals are divided have been already mentioned. In my hand I hold a compound plate, one half of it taken from a right-handed and the other from a left-handed crystal. Placing the plate in front of the polarizer, we turn one of the Nicols until the two halves of the plate show a common puce color. This yields an exceedingly sensitive means of rendering the action of a magnet upon light visible. By turning either the polarizer or the analyzer through the smallest angle, the uniformity of the color disappears, and the two halves of the quartz show different colors. The magnet also produces this effect. The puce-colored circle is now before you on the screen. (See Fig. 21 for the arrangement of the experiment. N is the nozzle of the lamp, H the first Nicol, Q the biquartz plate, L a lens, M the electromagnet, and P the second Nicol.) Exciting the magnet,

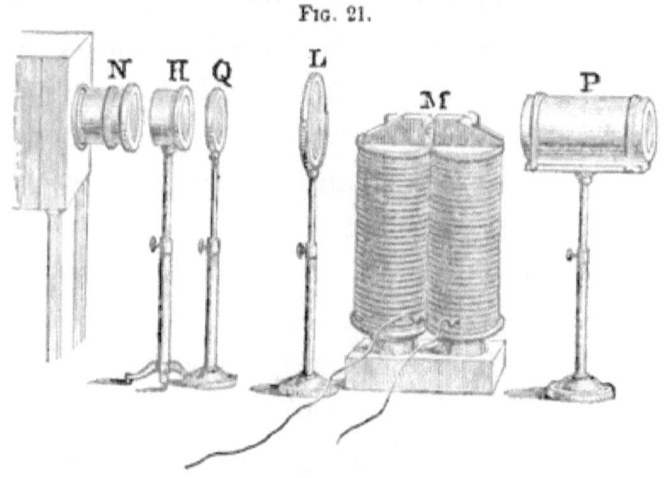

Fig. 21.

one half of the image becomes suddenly red, the other half green. Interrupting the current, the two colors fade away, and the primitive puce is restored. The action, moreover, depends upon the polarity of the magnet, or, in other words, on the direction of the current which surrounds the magnet. Reversing the current, the red and green reappear, but they have changed places. The red was formerly to the right, and the green to the left; the green is now to the right, and the red to the left. With the most exquisite ingenuity, Faraday analyzed all those actions and stated their laws. This experiment, however, long remained rather as a scientific curiosity than as a fruitful germ. That it would bear fruit of the highest importance, Faraday felt profoundly convinced, and recent researches are on the way to verify his conviction.

A few words more are necessary to complete our knowledge of the wonderful interaction between ponderable molecules and the ether interfused among them. Symmetry of molecular arrangement implies symmetry on the part of the ether; atomic dissymmetry, on the other hand, involves the dissymmetry of the ether, and, as a consequence, double refraction. In a certain class of crystals the structure is homogeneous, and such crystals produce no double refraction. In certain other crystals the molecules are ranged symmetrically round a certain line, and not around others. Along the former, therefore, the ray is undivided, while along all the others we have double refraction. Ice is a familiar example: it is built with perfect symmetry around the perpendiculars to the planes of freezing, and a ray sent through ice in this direction is not doubly refracted; whereas, in all other directions, it is. Iceland spar is another example of the same kind: its molecules are built symmetrically round the line uniting the two blunt angles of the rhomb. In this direction a ray suffers no double refraction, in all others it does. This direction of double refraction is called the *optic axis* of the crystal.

Hence, if a plate be cut from a crystal of Iceland spar perpendicular to the axis, all rays sent across this plate in the direction of the axis will produce but one image. But, the moment we deviate from the parallelism with the axis, double refraction sets in. If, therefore, a beam that has been rendered *conical* by a converging lens be sent through the spar so that the central ray of the cone passes along the axis, this ray only will escape double refraction. Each of the others will be divided into an ordinary and an extraordinary ray, the one moving more slowly through the crystal than the other; the one, therefore, retarded with reference to the other. Here, then, we have the conditions for interference, when the waves are reduced by the analyzer to a common plane. A highly beautiful and important source of chromatic phenomena is thus revealed. Placing the plate of spar between the crossed prisms, we have upon the screen a beautiful system of iris rings surrounding the end of the optic axis, the circular bands of color being intersected by a black cross. The arms of this cross are parallel to the two directions of vibration in the polarizer and analyzer. It is easy to see that those rays whose planes of vibration within the spar coincide with the plane of vibration of *either* prism, cannot get through *both*. This complete interception produces the arms of the cross. With monochromatic light the rings would be simply bright and black—the bright rings occurring at those thicknesses of the spar which cause the rays to conspire; the black rings at those thicknesses which cause them to quench each other. Here, however, as elsewhere, the different lengths of the light-waves give rise to iris-colors when white light is employed.

Besides the *regular* crystals which produce double refraction in no direction, and the *uniaxal* crystals which produce it in all directions but one, Brewster discovered that in a large class of crystals there are *two* directions in which double refraction does not take place. These are called *biaxal* crystals. When plates of these crystals, suitably cut, are placed between the polarizer and analyzer, the axes are seen surrounded, not by circles, but by curves of another order and of a perfectly definite mathematical character. Each band, as proved experimentally by Herschel, forms a *lemniscata;* but the experimental proof was here, as in numberless other cases, preceded by the deduction which showed that, according to the undulatory theory, the bands must possess this special character.

I have taken this somewhat wide range over polarization itself and over the phenomena exhibited by crystals in polarized light, in order to give you some notion of the firmness and completeness of the theory which grasps them all. Starting from the single assumption of transverse undulations, we first of all determine the wave-lengths, and find all the phenomena of color dependent on this element. The wave-lengths may be determined in many independent ways, and, when the lengths so determined are compared together, the strictest agreement is found to exist between them. We follow the ether into the most complicated cases of interaction between it and ordinary matter, "the theory is equal to them all. It makes not a single new physical hypothesis; but out of its original stock of principles it educes the counterparts of all that observation shows. It accounts for, explains, simplifies the most entangled cases; corrects known laws and facts; predicts and discloses unknown ones; becomes the guide of its former teacher Observation; and, enlightened by mechanical conceptions, acquires an insight which pierces through shape and color to force and cause."[1]

But, while I have thus endeavored to illustrate before you the power of the undulatory theory as a solver of all the difficulties of optics, do I therefore wish you to close your eyes to any evidence that may arise against it? By no means. You may urge, and justly urge, that a hundred years ago another theory was held by the most eminent men, and that, as the theory then held had to yield, the undulatory theory may have to yield also. This is perfectly logical; but let us understand the precise value of the argument. In similar language a person in the time of Newton, or even in our time, might reason thus: "Hipparchus and Ptolemy, and numbers of great men after them, believed that the earth was

[1] Whewell.

the centre of the solar system. But this deep-set theoretic notion had to give way, and the theory of gravitation may, in its turn, have to give way also." This is just as logical as the first argument. Wherein consists the strength of the theory of gravitation? Solely in its competence to account for all the phenomena of the solar system. Wherein consists the strength of the theory of undulation? Solely in its competence to disentangle and explain phenomena a hundred-fold more complex than those of the solar system. Be as skeptical, if you like, regarding the undulatory theory; but if your skepticism be philosophical, it will wrap the theory of gravitation in the same or greater doubt.[1]

[1] The only essay known to me on the Undulatory Theory, from the pen of an American writer, is an excellent one by President Barnard, published in the Smithsonian Report for 1862.

LECTURE V.

Range of Vision incommensurate with Range of Radiation: The Ultra-Violet Rays: Fluorescence: Rendering Invisible Rays visible: Vision not the only Sense appealed to by the Solar and Electric Beam: Heat of Beam: Combustion by Total Beam at the Foci of Mirrors and Lenses: Combustion through Ice-Lens: Ignition of Diamond: Search for the Rays here effective: Sir William Herschel's Discovery of Dark Solar Rays: Invisible Rays the Basis of the Visible: Detachment by a Ray-Filter of the Invisible Rays from the Visible: Combustion at Dark Foci: Conversion of Heat-Rays into Light-Rays: Calorescence: Part played in Nature by Dark Rays: Identity of Light and Radiant Heat: Invisible Images: Reflection, Refraction, Plane Polarization, Depolarization, Circular Polarization, Double Refraction, and Magnetization of Radiant Heat.

The first question that we have to consider tonight is this: Is the eye, as an organ of vision, commensurate with the whole range of solar radiation—is it capable of receiving visual impressions from all the rays emitted by the sun? The answer is negative. If we allowed ourselves to accept for a moment that notion of gradual growth, amelioration, and ascension, implied by the term *evolution*, we might fairly conclude that there are stores of visual impressions awaiting man far greater than those of which he is now in possession. For example, here beyond the extreme violet of the spectrum there is a vast efflux of rays which are totally useless as regards our present powers of vision. But these ultra-violet waves, though incompetent to awaken the optic nerve, can so shake the molecules of certain compound substances as to effect their decomposition. The grandest example of the chemical action of light, with which my friend Dr. Draper has so indissolubly associated his name, is that of the decomposition of carbonic acid in the leaves of plants. All photography is founded on such actions. There are substances on which the ultra-violet waves exert a special decomposing power; and, by permitting the invisible spectrum to fall upon surfaces prepared with such substances, we reveal both the existence and the extent of the ultra-violet spectrum.

This mode of exhibiting the action of the ultra-violet rays has been long known; indeed, Thomas Young photographed the ultra-violet rings of Newton. We have now to demonstrate their presence in another way. As a general rule, bodies transmit light or absorb it, but there is a third case in which the light falling upon the body is neither transmitted nor absorbed, but converted into light of another kind. Professor Stokes, the occupant of the chair of Newton in the University of Cambridge, one of those original workers who, though not widely known beyond scientific circles, really constitute the core of science, has demonstrated this change of one kind of light into another, and has pushed his experiments so far as to render the invisible rays visible.

A long list of substances examined by Stokes when

excited by the invisible ultra-violet waves, have been proved to emit *light*. You know the rate of vibration corresponding to the extreme violet of the spectrum; you are aware that, to produce the impression of this color, the retina is struck 789 millions of millions of times in a second. At this point, the retina ceases to be useful as an organ of vision, for, though struck by waves of more rapid recurrence, they are incompetent to awaken the sensation of light. But, when such non-visual waves are caused to impinge upon the molecules of certain substances—on those of sulphate of quinine, for example—they compel those molecules, or their constituent atoms, to vibrate; and the peculiarity is, that the vibrations thus set up are *of slower period* than those of the exciting waves. By this lowering of the rate of vibration through the intermediation of the sulphate of quinine, the invisible rays are rendered visible. Here we have our spectrum, and beyond the violet I place this prepared paper. The spectrum is immediately elongated by the generation of new light beyond the extreme violet. President Morton has recently succeeded in discovering a substance of great sensibility which he has named *Thallene*, and he has been good enough to favor me with some paper saturated with a solution of this substance. It causes a very striking elongation of the spectrum, the new light generated being of peculiar brilliancy. To this change of the rays from a higher to a lower refrangibility, Stokes has given the name of *Fluorescence.*

By means of a deeply-colored violet glass, we cut off almost the whole of the *light* of our electric beam; but this glass is peculiarly transparent to the violet and ultra-violet rays. The violet beam now crosses a large jar filled with water. Into it I pour a solution of sulphate of quinine: opaque clouds, to all appearance, instantly tumble downwards. But these are not clouds: there is nothing precipitated here: the observed action is an action of *molecules,* not of *particles.* The medium before you is not a turbid medium, for, when you look through it at a luminous surface, it is perfectly clear. If we paint upon a piece of paper a flower or a bouquet with the sulphate of quinine, and expose it to the full beam, scarcely any thing is seen. But on interposing the violet glass, the design instantly flashes forth in strong contrast with the deep surrounding violet. Here is such a design prepared for me by President Morton with his thallene: placed in the violet light it exhibits a peculiarly vivid and beautiful fluorescence. From the experiments of Dr. Bence Jones, it would seem that there is some substance in the human body resembling the sulphate of quinine, which causes all the tissues of the body to be more or less fluorescent. The crystalline lens of the eye exhibits the effect in a very striking manner. When I plunge my eye into this violet beam, I am conscious of a whitish-blue shimmer filling the space before me. This is caused by fluorescent light generated in the eye itself; looked at from without, the crystalline lens at the same time gleams vividly.

But the waves from our incandescent carbon-points appeal to another sense than that of vision. They not only produce light as a sensation; they also produce heat. The magnified image of the carbon-points is now upon the screen; and with a suitable instrument the heating power of that image might be demonstrated. Here, however, the heat is spread over too large an area to be intense. By pushing out the lens and causing a movable screen to approach our lamp, the image becomes smaller and smaller: the rays become more concentrated, until finally they are able to pierce black paper with a burning ring. Rendering the beam parallel, and receiving it upon a concave mirror, the rays are brought to a focus; and paper placed at the focus is caused to smoke and burn. This may be done by our common camera with its lens, and by a concave mirror of very moderate power.

We will now adopt stronger measures with the radiation from the electric lamp. In this camera of blackened tin is placed a lamp, in all particulars similar to those already employed. But, instead of gathering up the rays from the carbon-points by a condensing lens placed in front of them, we gather them up by a concave mirror, silvered in front, and placed behind the carbons. By this mirror we can cause the rays to issue through the orifice in front, either parallel or convergent. They are now parallel, and therefore to a certain extent diffused. We place a convex lens in the path of the beam; the light is converged to a focus, and at that focus you see that paper is not only pierced and a burning ring formed, but that it is instantly set ablaze. Many metals may be burned up in the same way. In our first lecture the combustibility of zinc was mentioned. Placing a strip of sheet-zinc at this focus, it is instantly ignited and burns with its characteristic purple flame. (In the annexed figure $m\ m'$ represents

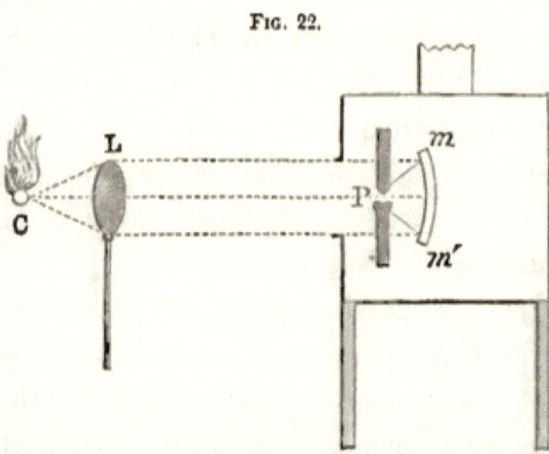

Fig. 22.

the concave mirror, L the lens, at the focus C of which combustion is effected.) Dr. Scoresby succeeded in exploding gunpowder by the sun's rays converged by large lenses of ice; the same effect may be produced with a small lens, and with a terrestrial source of heat. In an iron mould we have fashioned this beautiful lens of transparent ice. At the focus of the lens I place a bit of black paper, with a little gun-cotton folded up within it. The paper ignites and the cotton explodes. Strange, is it not, that the beam should possess such heating power after having passed through so cold a substance?

In this experiment, you observe that, before the beam reaches the ice-lens, it has passed through a glass cell containing water. The beam is thus sifted of constituents, which, if permitted to fall upon the lens, would injure its surface, and blur the focus. And this leads me to say an anticipatory word regarding transparency. In our first lecture we entered fully into the production of colors by absorption, and we spoke repeatedly of the quenching of the rays of light. Did this mean that the light was altogether annihilated? By no means. It was simply so lowered in refrangibility as to escape the visual range. *It was converted into heat.* Our red ribbon in the green of the spectrum quenched the green, but if suitably examined its temperature would have been found raised. Our green ribbon in the red of the spectrum quenched the red, but its temperature at the same time was augmented to a degree exactly equivalent to the light extinguished. Our black ribbon, when passed through the spectrum, was found competent to quench all its colors; but at every stage of its progress an amount of heat was generated in the ribbon exactly equivalent to the light lost. *It is only when absorption takes place that heat is thus produced;* and heat is always a result of absorption.

Examine this water, then, in front of the lamp, after the beam has passed a little time through it: it is sensibly warm, and, if permitted to remain there long enough, it may be made to boil. This is due to the absorption by the water of a portion of the electric beam. But a certain portion passes through unabsorbed, and does not at all contribute to the heating of the water. Now, ice is also transparent to the latter portion, and therefore is not melted by it; hence, by employing this particular portion of the beam, we are able to keep our lens intact, and to produce by means of it a sharply-defined focus. Placed at that focus, black paper instantly burns, because the black paper absorbs the light which had passed through the ice-lens without absorption. In a subsequent lecture we shall endeavor to penetrate further into the physical meaning of these and other similar actions. I may add to these illustrations of heating power, the ignition of a diamond in oxygen, by the concentrated beam of the electric lamp. The diamond, surrounded by a hood of platinum to lessen the chilling due to convection, is exposed at the focus. It is rapidly raised to a white heat, and when removed from the focus continues to glow like a star.

Placed in the path of the beam issuing from our lamp is a cell with glass sides containing a solution of alum. All the *light* of the beam passes through this solution. The beam is received on a powerfully converging mirror silvered in front, and is brought to a focus by the mirror. You can see the conical beam of reflected light tracking itself through the dust of the room. I place at the focus a scrap of white paper: it glows there with dazzling brightness, but it is not even charred. On removing the alum-cell, however, the paper instantly inflames. There must, therefore, be something in this beam besides its light. The *light* is not absorbed by the white paper, and therefore does not burn the paper; but there is something over and above the light which *is* absorbed and which provokes combustion. What is this something?

In the year 1800 Sir William Herschel passed a thermometer through the various colors of the solar spectrum, and marked the rise of temperature corresponding to each color. He found the heating effect to augment from the violet to the red; he did not, however, stop at the red, but pushed his thermometer into the dark space beyond it. Here he found the temperature actually higher than in any part of the visible spectrum. By this important observation, he proved that the sun emitted dark heat-rays which are entirely unfit for the purposes of vision. The subject was subsequently taken up by Seebeck, Melloni, Müller, and others, and within the last few years it has been found capable of unexpected expansions and applications. A method has been devised whereby the solar or electric beam can be so *filtered* as to detach from it and preserve intact this invisible ultra-red emission, while the visible and ultra-violet emissions are wholly intercepted. We are thus enabled to operate at will upon the purely ultra-red waves.

In the heating of solid bodies to incandescence this non-visual emission is the necessary basis of the visual. A platinum wire is stretched in front of the table, and through it an electric current flows. It is warmed by the current, and may be felt to be warm by the hand; it also emits waves of heat, but no light. Augmenting the strength of the current, the wire becomes hotter; it finally glows with a sober red light. At this point Dr. Draper many years ago began an interesting investigation. He employed a voltaic current to heat his platinum, and he studied by means of a prism the successive introduction of the colors of the spectrum. His first color, as here, was red; then came orange, then yellow, then green, and lastly all the shades of blue. Thus as the temperature of the platinum was gradually augmented, the atoms were caused to vibrate more rapidly, shorter waves were thus produced, until finally he obtained the waves corresponding to the entire spectrum. As each successive color was introduced, the colors preceding it became more vivid. Now, the vividness, or intensity of light, like that of sound, depends, not upon the length of the wave, but on the amplitude of the vibration. Hence, as the red grew more intense as the more refrangible colors were introduced, we are forced to conclude that, side by side with the introduction of the shorter waves, we had an augmentation of the amplitude of the longer ones.

These remarks apply, not only to the visible emission examined by Dr. Draper, but to the invisible emission which preceded the appearance of any light. In the emission from the white-hot platinum wire now before you the very waves exist with which we started, only their intensity has been increased a thousand-fold by the augmentation of temperature necessary to the production of this white light. Both effects are bound together: in an incandescent solid, or in a molten solid, you cannot have the shorter waves without this intensification of the longer ones. A sun is possible only on these conditions; hence Sir William Herschel's discovery of the invisible ultra-red solar emission.

The invisible heat, emitted both by dark bodies and by luminous ones, flies through space with the velocity of light, and is called *radiant heat*. Now, radiant heat may be made a subtle and powerful explorer of molecular condition, and of late years it has given a new significance to the act of chemical combination. Take, for example, the air we breathe. It is a mixture of oxygen and nitrogen; and with regard to radiant heat it behaves like a vacuum, being incompetent to absorb it in any sensible degree. But permit the same two gases to unite chemically; without any augmentation of the quantity of matter, without altering the gaseous condition, without interfering in any way with the *transparency*, of the gas, the act of chemical union is accompanied by an enormous diminution of its *diathermancy*, or perviousness to radiant heat. The researches which established this result also proved the elementary gases generally to be highly transparent to radiant heat. This, again, led to the proof of the diathermancy of elementary *liquids*, like bromine, and of *solutions* of the elements sulphur, phosphorus, and iodine. A spectrum is now before you, and you notice that this transparent bisulphide of carbon has no effect upon the colors. Dropping into the liquid a few flakes of iodine, you see the middle of the spectrum cut away. By augmenting the quantity of iodine, we invade the entire spectrum, and finally cut it off altogether. Now, the iodine which proves itself thus hostile to the light is perfectly transparent to the ultra-red emission with which we have now to deal. It, therefore, is to be our ray-filter.

Placing the alum-cell again in front of the electric lamp, we assure ourselves as before of the utter inability of the concentrated *light* to fire white paper. By introducing a cell containing the solution of iodine, the light is entirely cut off. On removing the alum-cell, the paper at the dark focus is instantly set on fire, black paper is more absorbent than white for these ultra-red rays; and the consequence is, that with it the suddenness and vigor of the combustion are augmented. Zinc is burnt up at the same place, while magnesium ribbon bursts into vivid combustion. A sheet of platinized platinum placed at the focus is heated to whiteness. Looked at through a prism, the white-hot platinum yields all the colors of the spectrum. Before impinging upon the platinum, the waves were of too slow recurrence to awaken vision; by the atoms of the platinum, these long and sluggish waves are in part broken up into shorter ones, being thus brought within the visual range. At the other end of the spectrum, Stokes, by the interposition of suitable substances, *lowered* the refrangibility so as to render the non-visual rays visual, and to this change he gave the name of *Fluorescence*. Here, by the intervention of the platinum, the refrangibility is *raised*, so as to render the non-visual visual, and to this change we give the name of *Calorescence*.

At the perfectly invisible focus where these effects are produced, the air may be as cold as ice. Air, as already stated, does not absorb the radiant heat, and is therefore not warmed by it. Place at the focus the most sensitive air-thermometer: it is not affected by the heat. Nothing could more forcibly illustrate the isolation, if I may use the term, of the luminiferous ether from the air. The wave-motion of the one is heaped up, without sensible effect upon the other. I may add that, with suitable precautions, the eye may be placed in a focus competent to heat platinum to vivid redness, without experiencing any damage, or the slightest sensation either of light or heat.

These ultra-red rays play a most important part in Nature. I remove the iodine filter, and concentrate the total beam. A test-tube containing water is placed at the focus: it immediately begins to sputter, and in a minute or two it *boils*. What boils it? Placing the alum solution in front of the lamp, the boiling instantly ceases. Now, the alum is pervious to all the luminous rays; hence it cannot be these rays that caused the boiling. I now introduce the iodine, and remove the alum; vigorous ebullition immediately recommences. So that we here fix upon the invisible ultra-red rays the heating of the water. We are enabled now to understand the momentous part played by these rays in Nature. It is to them that we owe the warming and the consequent evaporation of the tropical ocean; it is to them, therefore, that we owe our rains and snows. They are absorbed close to the surface of the ocean, and warm the superficial water, while the luminous rays plunge to great depths without producing any sensible effect. Further, here is a large flask containing a freezing mixture. The aqueous vapor of the air has been condensed and frozen on the flask, which is now covered with a white fur. Introducing the alum-cell, we place the coating of hoar-frost at the intensely luminous focus; not a spicula of the frost is melted. Introducing the iodine-cell, and removing the alum, a broad space of the frozen coating is instantly removed. Hence we infer that the ice which feeds the Rhone, the Rhine, and other rivers which have glaciers for their sources, is released from its imprisonment upon the mountains by the invisible ultra-red rays of the sun.

The growth of science is organic. The end of to-day becomes to-morrow the means to a remoter end. Every new discovery is immediately made the basis of other discoveries, or of new methods of investigation. About fifty years ago, Œrsted, of Copenhagen, discovered the deflection of a magnetic needle by an electric current; and Thomas Seebeck, of Berlin, discovered that electric currents might be derived from heat. Soon afterwards these discoveries were turned to account by Nobili and Melloni in the construction of an apparatus which has vastly augmented our knowledge of radiant heat. The instrument is here. It is called a *thermo-electric pile;* and it consists of thin bars of bismuth and antimony, soldered together in pairs at their ends, but separated from each other elsewhere. From the ends of this "pile" wires pass to a coil of covered wire, within and above which are suspended two magnetic needles joined to a rigid system, and carefully defended from currents of air. The heat, then, acting on the pile, produces an electric current; the current, passing through the coil deflects the needles, and the magnitude of the deflection may be made a measure of the heat. The upper needle moves over a graduated dial far too small to be seen. It is now, however, strongly illuminated. Above it is a lens which, if permitted, would form an image of the needle and dial upon the ceiling, where, however, it could not be conveniently seen. The beam is therefore received upon a looking-glass, placed at the proper angle, which throws the image upon the screen. In this way the motions of this small needle may be made visible to you all.

The delicacy of this instrument is such that in a room like this it is exceedingly difficult to work with it. My assistant stands several feet off. I turn the pile towards him: the heat from his face, even at this distance, produces a deflection of 90°. I turn the instrument towards a distant wall, which I judge to be a little below the average temperature of the room. The needle descends and passes to the other side of zero, declaring by this negative deflection that the pile feels the chill of the wall. Possessed of this instrument, of our ray-filter, and of our large Nicol prisms, we are in a condition to investigate a subject of great philosophical interest, and which long engaged the attention of some of our foremost scientific workers, Forbes being the first successful one—the substantial *identity of light and radiant heat.*

That they are identical in *all* respects cannot of course be the case, for if they were they would act in the same manner upon all instruments, the *eye* included. The identity meant is such as subsists between one color and another, causing them to behave alike as regards reflection, refraction, double refraction, and polarization. As regards reflection, we may employ the looking-glass used in our first lecture. Marking any point in the track of the reflected beam, and cutting off the light by the iodine, on placing the pile at the marked point, the needle immediately starts aside. This is true for every position of the mirror. So that both for light and heat the same law of reflection holds good; for both of them also the angular velocity of the reflected beam is twice that of the reflecting mirror. Receiving the beam on a concave mirror, it is gathered up into a cone of reflected light; marking the apex of the cone, and cutting off the light, a moment's exposure of the pile at the marked point produces a violent deflection of the needle. (See Fig. 23, where $m\ m$ is the mirror, P the pile, and T the opaque solution.)

This beam of light now enters a right-angled prism and is reflected at the hypothenuse, in a direction perpendicular to its former one. The reflection here is *total*. Cutting off the light, we prove the reflection of the heat to be total also. The formation

FIG. 23.

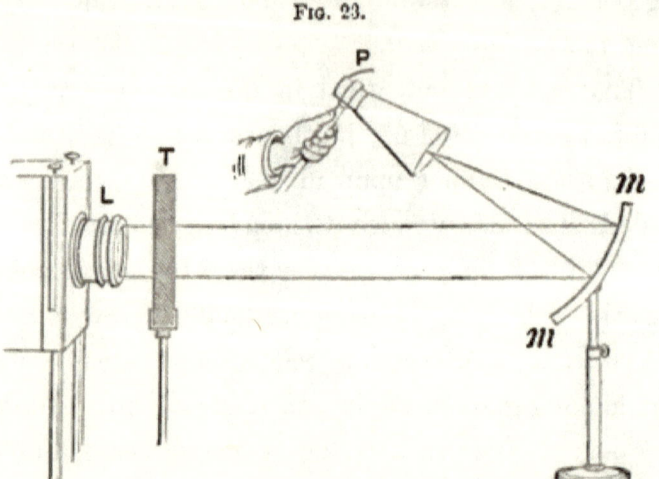

of invisible images by lenses and mirrors may also be demonstrated. Concentrating the beam, and cutting off the light, at the dark focus the carbon-points *burn* their images through a sheet of black paper. Placing a sheet of platinized platinum at the focus, when the concentration is strong an incandescent image of the points is immediately stamped upon the platinum.

And now for polarization and its attendant phenomena. Crossing our two Nicol prisms, B, C, Fig. 24 and placing our pile D behind the analyzer, neither heat nor light reaches it; the needle remains undeflected. Introducing the iodine, the slightest turning of either prism causes the heat to pass, and to announce itself by the deflection of the needle. Like light, therefore, heat is polarized. Crossing the Nicols again, the heat is intercepted and the needle returns

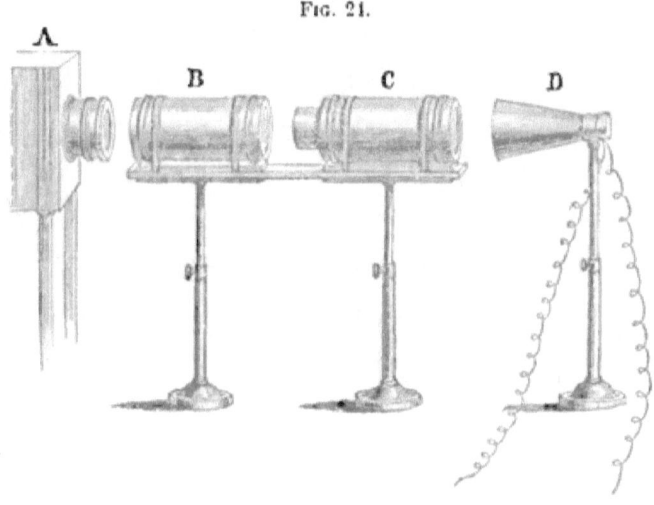

Fig. 21.

to zero. Plunging into the dark space between the prisms our plate of mica, the needle instantly starts off, showing that the mica acts upon the heat as it did upon the light: we have in both cases the same resolution and recompounding of vibrations. Removing the mica, the needle falls to zero; but, on introducing a plate of quartz between the prisms, the consequent deflection declares the circular polarization of the heat. For double refraction it is necessary that our images should not be too large and diluted: here are the two disks produced by the splitting of the beam in Iceland spar. Marking the positions of the disks and cutting off the light, the pile finds in its places two heat-images. The needle now stands near 90°, and, on turning the spar, the deflection remains constant. Transferring the pile to the other image, the deflection of 90° is maintained; but on turning the spar the needle now falls to zero. The reason is manifest. Permitting the light to pass, we find the luminous disk at some distance from the pile. We are dealing, in fact, with the *extraordinary beam* which rotates round the ordinary. So that for heat as well as for light we have double refraction, and also an ordinary and extraordinary ray. (In the adjacent figure, which shows the ex-

Fig. 25.

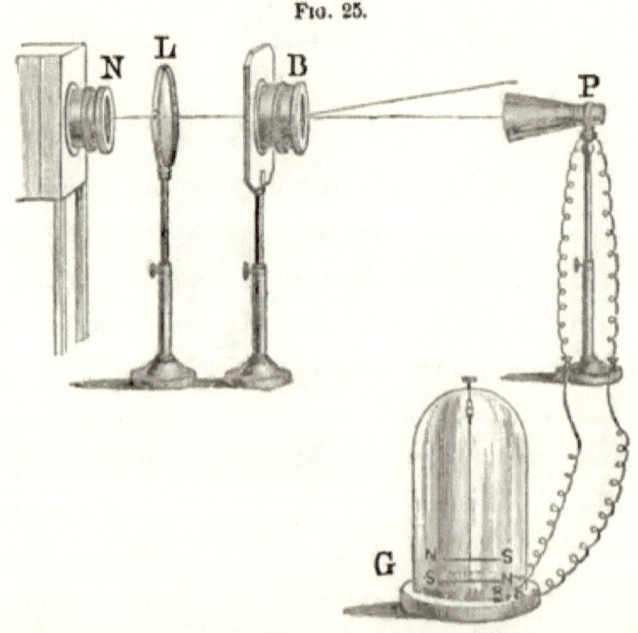

perimental arrangement, N is the nozzle of the electric lamp, L a converging lens, B the birefracting spar, and P the thermo-electric pile.)

DOUBLE REFRACTION OF HEAT.

If time permitted, we might finish this series of demonstrations by magnetizing a ray of heat as we magnetized a ray of light.

We have finally to determine the position and magnitude of the invisible radiation which produces these results. For this purpose we employ a particular form of the thermo-electric pile. Its face is a rectangle, which by movable side-pieces can be rendered as narrow as desirable. Throwing a concentrated spectrum upon a *small* screen, by means of an endless screw, we move this rectangular pile through the entire spectrum. Its surface is blackened so that it absorbs all the light incident upon it, converting it into heat, and thus enabling it to declare its power by the deflection of the magnetic needle.

When this instrument is brought to the violet end of the spectrum, the heat is found to be almost insensible. As the pile gradually moves from the violet towards the red, it encounters a gradually-augmenting heat. The red itself possesses the highest heating power of all the colors of the spectrum. Pushing the pile into the dark space beyond the red, the heat rises suddenly in intensity, and, at some distance beyond the red, attains a maximum. From this point the heat falls somewhat more rapidly than it rose, and afterwards gradually fades away. Drawing an horizontal line to represent the length of the spectrum, and erecting along it, at various points, perpendiculars propor-

red.

Spectrum of Electric Light.

tional in length to the heat existing at those points, we obtain a curve which exhibits the distribution of heat in our spectrum. It is represented in the adjacent figure. Beginning at the blue, the curve rises, at first very gradually; then as it approaches the red more rapidly, the line C D representing the strength of the extreme red radiation. Beyond the red it shoots upwards in a steep and massive peak to B, whence it falls, rapidly for a time, and afterwards gradually fading from the perception of the pile. This figure is the result of more than twelve careful series of measurements, for each of which the curve was constructed. On superposing all these curves, a satisfactory agreement was found to exist between them. So that it may safely be concluded that the areas of the dark and white spaces respectively represent the relative energies of the visible and invisible radiation. The one is 7.7 times the other.

But in verification, as already stated, consists the strength of science. Determining in the first place the total emission from the electric lamp; then by means of the iodine filter determining the ultra-red emission; the difference between both gives the luminous emission. In this way, it was found that the energy of the invisible emission is eight times that of the visible. No two methods could be more opposed to each other, and hardly any two results could better harmonize. I think, therefore, you may rely upon the accuracy of the distribution of heat here assigned to the prismatic spectrum of the electric light. There is nothing vague in the mode of investigation, nor doubtful in its conclusions.

LECTURE VI.

Principles of Spectrum Analysis: Solar Chemistry: Summary and Conclusion.

WE have employed, as our source of light in these lectures, the ends of two rods of coke rendered incandescent by electricity. Coke is particularly suitable for this purpose, because it can bear intense heat without fusion or vaporization. It is also black, which helps the light; for, other circumstances being equal, as shown experimentally by Balfour Stewart, the blacker the body the brighter will be its light when incandescent. Still, refractory as carbon is, if we closely examined our voltaic arc, or stream of light between the carbon-points, we should find there incandescent carbon-vapor. We might also detach the light of this vapor from the more dazzling light of the solid points, and obtain its spectrum. This would be not only less brilliant, but of a totally different character from the spectra that we have already seen. Instead of being an unbroken succession of colors from red to violet, the carbon-vapor would yield a few bands of color with spaces of darkness between them.

What is true of the carbon is true in a still more striking degree of the metals, the most refractory of which can be fused, boiled, and reduced to vapor by the electric current. From the incandescent vapor the light, as a general rule, flashes in groups of rays of definite degrees of refrangibility, spaces existing between group and group, which are unfilled by rays of any kind. But the contemplation of the facts will render this subject more intelligible than words can make it. Within the camera is now placed a cylinder of carbon hollowed out at the top to receive a bit of metal; in the hollow is placed a fragment of the metal thallium, and now you see the arc of incandescent thallium-vapor upon the screen. It is of a beautiful green color. What is the meaning of that green? We answer the question by subjecting the light to prismatic analysis. Here you have its spectrum, consisting of a single refracted band. Light of one degree of refrangibility, and that corresponding to green, is emitted by the thallium-vapor.

We will now remove the thallium and put a bit of silver in its place. The arc of silver is not to be distinguished from that of thallium; it is not only green, like the thallium-vapor, but the same shade of green. Are they, then, alike? Prismatic analysis enables us to answer the question. It is perfectly impossible to confound the spectrum of incandescent silver vapor with that of thallium. Here are two green bands instead of one. Adding to the silver in our camera a bit of thallium, we obtain the light of both metals, and you see that the green of the thallium lies midway between the two greens of the silver. Hence this similarity of color.

But you observe another interesting fact. The thallium band is now far brighter than the silver bands; indeed, the latter have wonderfully degenerated since the bit of thallium was put in. They are not at all so bright as they were at first, and for a reason worth knowing. It is the *resistance* offered to the passage of the electric current from carbon to carbon that calls forth the power of the current to produce heat. If the resistance were materially lessened, the heat would be materially lessened; and, if all resistance were abolished, there would be no heat at all. Now, thallium is a much more fusible and vaporizable metal than silver; and its vapor facilitates the passage of the current to such a degree as to render it almost incompetent to vaporize the silver. But the thallium is gradually consumed; its vapor diminishes, the resistance rises, until finally you see the two silver bands as brilliant as they were at first. The three bands of the two metals are now of the same sensible brightness.

We have in these bands a perfectly unalterable characteristic of these two metals. You never get other bands than these two green ones from the silver, never other than the single green band from the thallium, never other than the three green bands from the mixture of both metals. Every known metal has its bands, and in no known case are the bands of two different metals alike. Hence these spectra may be made a test for the presence or absence of any particular metal. If we pass from the metals to their alloys, we find no confusion. Copper gives us green bands, zinc gives us blue and red bands; brass, an alloy of copper and zinc, gives us the bands of both metals, perfectly unaltered in position or character.

But we are not confined to the metals; the *salts* of these metals yield the bands of the metals. Chemical union is ruptured by a sufficiently high heat, the vapor of the metal is set free and yields its characteristic bands. The chlorides of the metals are particularly suitable for experiments of this character. Common salt, for example, is a compound of chlorine and sodium; in the electric lamp it yields the spectrum of the metal sodium. The chlorides of lithium and of strontium yield in like manner the bands of these metals. When, therefore, Bunsen and Kirchhoff, the celebrated founders of *spectrum analysis*, after having established by an exhaustive examination the spectra of all known substances, discovered a spectrum containing bands different from any known bands, they immediately inferred the existence of a new metal. They were operating at the time upon a residue obtained by evaporating one of the mineral waters of Germany. In that water they knew the new metal was concealed, but vast quantities of it had to be evaporated before a residue could be obtained sufficiently large to enable ordinary chemistry to grapple with the metal. But they hunted it down, and it now stands among chemical substances as the metal *Rubidium*. They subsequently discovered a second metal, which they called *Cæsium*. Thus, having first placed spectrum analysis on a safe foundation, they demonstrated its capacity as an agent of discovery. Soon afterwards Mr. Crookes, pursuing this same method, obtained the salts of the thallium which yielded that bright monochromatic green band. The metal itself was first isolated by a French chemist.

All this relates to chemical discovery upon earth, where the materials are in our own hands. But Kirchhoff showed how spectrum analysis might be applied to the investigation of the sun and stars, and on his way to this result he solved a problem which had been long an enigma to natural philosophers. A spectrum is *pure* in which the colors do not overlap each other. We purify the spectrum by making our slit narrow and by augmenting the number of our prisms. When a pure spectrum of the sun has been obtained in this way it is found to be furrowed by innumerable dark lines. Four of them were first seen by Dr. Wollaston, but they were afterwards multiplied and measured by Fraunhofer with such masterly skill that they are now universally known as Fraunhofer's

Fig. 27.

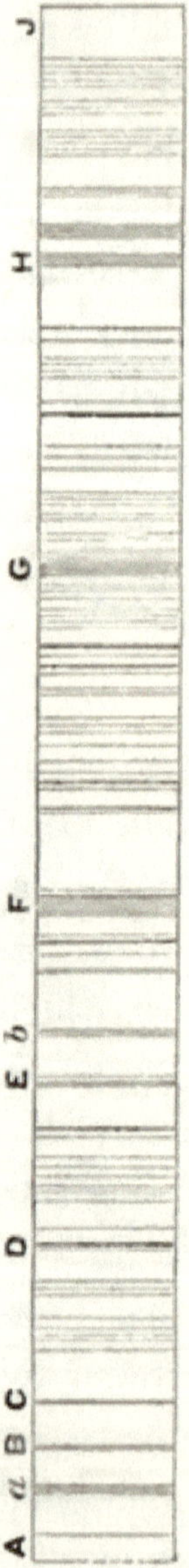

lines. To give an explanation of these lines was, as I have said, a problem which long challenged the attention of philosophers. (The principal lines are lettered according to Fraunhofer in the annexed sketch of the solar spectrum. A, it may be stated, stands near the extreme red, and J near the extreme violet.)

Now, Kirchhoff had made thoroughly clear to his mind the principles which linked together the *emission* of light and the *absorption* of light; he had proved their inseparability for each particular kind of light and heat. He had proved, for every specific ray of the spectrum, the doctrine that the body emitting any ray absorbed with special energy a ray of the same refrangibility. Consider, then, the effect of knowledge, such as you now possess, upon a mind prepared like that of Kirchhoff. We have seen the incandescent vapors of metals emitting definite groups of rays; according to Kirchhoff's principle, those vapors, if crossed by solar light, ought to absorb rays of the same refrangibility as those which they emit. He proved this to be the case; he was able, by the interposition of a vapor, to cut out of the solar spectrum the band corresponding in color to that vapor. Now, the sun possesses a photosphere, or vaporous envelope—doubtless mixed with violently agitated clouds—and Kirchhoff saw that the powerful rays coming from the solid, or the molten nucleus of the sun, must be intercepted by this vapor. One dark band of Fraunhofer, for example, occurs in the yellow of the spectrum. Sodium vapor is demonstrably competent to produce that dark band; hence Kirchhoff inferred the existence of sodium-vapor in the atmosphere of the sun. In the case of metals, which emit a large number of bands, the absolute coincidence of every bright band of the metal, with a dark Fraunhofer line, raises to the highest degree of certainty the inference that the metal is present in the atmosphere of the sun. In this way *solar chemistry* was founded on spectrum analysis.

But let me not skim so lightly over this great subject. I have spoken of emission and absorption, and of the link that binds them. Let me endeavor to make plain to you, through the analogy of sound, their physical meaning. I draw a fiddle-bow across this tuning-fork, and it immediately fills the room with a musical sound; this may be regarded as the *radiation* or *emission* of sound from the fork. A few days ago, on sounding this fork, I noticed that, when its vibrations were quenched, the sound seemed to be continued, though more feebly. The sound appeared to come from under a distant table, where stood a number of tuning-forks of different sizes and rates of vibration. One of these, and one only, had been started by the fork, and it was one whose rate of vibration was the same as that of the fork which started it. This is an instance of the *absorption* of the sound of one fork by another. Placing two forks near each other, sweeping the bow over one of them, and then quenching the agitated fork, the other continues to sound. Placing a cent-piece on each prong of one of the forks, we destroy its perfect synchronism with the other, and then no communication of sound from the one to the other is possible.

I will now do with *light* what has been here done with sound. Placing a tin spoon containing sodium in a Bunsen's flame, we obtain this intensely-yellow light, which corresponds in refrangibility with the yellow band of the spectrum. Like our tuning-fork, it emits waves of a special period. I will send the white light from our lamp through that flame, and prove before you that the yellow flame intercepts the yellow of the spectrum S S, Fig. 28, in other words, absorbs waves of the same period as its own, thus producing, to all intents and purposes, a dark Fraunhofer's band in the place of the yellow. (A Bunsen's flame contained within the chimney C is placed in front of the lamp L. The tin spoon with its pellet of sodium is plunged into the flame. Vivid combustion soon sets in; and, when it does, the yellow of the spectrum, at D, is furrowed by a dark band. Withdrawing and introducing the sodium-flame in rapid succession, the sudden disappearance and reappearance of the strip of darkness are observed.)

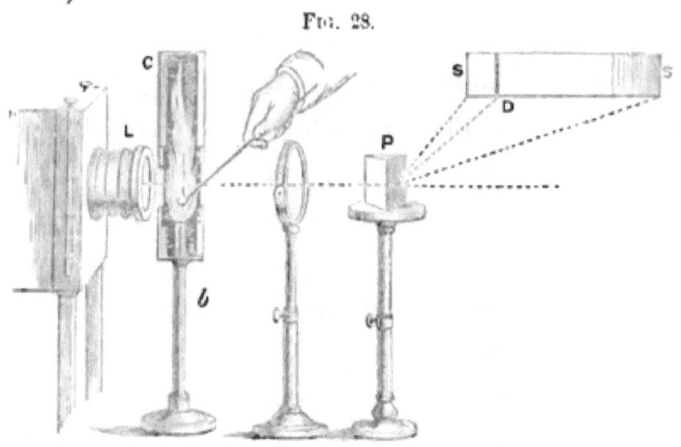

Fig. 28.

Mentally, as well as physically, every age of the world is the outgrowth and offspring of all preceding ages. Science proves itself to be a genuine product of Nature by growing according to this law. We have no solution of continuity here. Every great discovery has been duly prepared for in two ways: first, by other discoveries which form its prelude; and, secondly, through the sharpening, by exercise, of the intellectual instrument itself. Thus Ptolemy grew out of Hipparchus, Copernicus out of both, Kepler out of all three, and Newton out of all the four. Newton did not rise suddenly from the sea-level of the intellect to his amazing elevation. At the time that he appeared, the table-land of knowledge was already high. He juts, it is true, above the table-land, as a massive peak; still he is supported by it, and a great part of his absolute height was the height of humanity in his time. It is thus with the discoveries of Kirchhoff. Much had been previously accomplished; this he mastered, and then by the force of individual genius went beyond it. He replaced uncertainty by certainty, vagueness by definiteness, confusion by order; and I do not think that Newton has a surer claim to the discoveries that have made his name immortal, than Kirchhoff has to the credit of gathering up the fragmentary knowledge of his time, of vastly extending it, and of infusing into it the life of great principles. Splendid results have since been obtained, with which the names of Janssen, Huggins, Lockyer, Respighi, Young, and others, are honorably associated, but, splendid as they are, they are but the sequel and application of the principles established in his Heidelberg laboratory by the celebrated German investigator.

SUMMARY AND CONCLUSION.

My desire in these lectures has been to show you, with as little breach of continuity as possible, the past growth and present aspect of a department of science, in which have labored some of the greatest intellects the world has ever seen. My friend Professor Henry, in introducing me at Washington, spoke of me as an apostle; but the only apostolate that I intended to fulfil was to place, in plain words, my subject before you, and to permit its own intrinsic attractions to act upon your minds. In the way of experiment, I have tried to give you the best which, under the circumstances, could be provided; but I have sought to confer on each experiment a distinct intellectual value, for experiments ought to be the representatives and expositors of thought—a language addressed to the eye as spoken words are to the ear. In association with its context, nothing is more impressive or instructive than a fit experiment; but, apart from its context, it rather suits the conjurer's purpose of surprise than that purpose of education which ought to be the ruling motive of the scientific man.

And now a brief summary of our work will not be out of place. Our present mastery over the laws and phenomena of light has its origin in the desire of man to *know*. We have seen the ancients busy with this problem, but, like a child who uses his arms aimlessly for want of the necessary muscular exercise, so these early men speculated vaguely and confusedly regarding light, not having as yet the discipline needed to give clearness to their insight, and firmness to their grasp of principles. They assured themselves of the rectilineal propagation of light, and that the angle of incidence was equal to the angle of reflection. For more than a thousand years—I might say, indeed, for more than fifteen hundred years subsequently—the scientific intellect appears as if smitten with paralysis, the fact being that, during this time, the mental force, which might have run in the direction of science, was diverted into other directions.

The course of investigation as regards light was resumed in 1100 by an Arabian philosopher named Alhazan. Then it was taken up in succession by Roger Bacon, Vitellio, and Kepler. These men, though failing to detect the principle which ruled the facts, kept the fire of investigation constantly burning. Then came the fundamental discovery of Snell, that cornerstone of optics, as I have already called it, and immediately afterward we have the application by Descartes of Snell's discovery to the explanation of the rainbow. Then came Newton's crowning experiments on the anyalysis and synthesis of white light, by which it was proved to be compounded of various kinds of light of different degrees of refrangibility.

In 1676 an impulse was given to optics by astronomy. In that year Olaf Roemer, a learned Dane, was engaged at the Observatory of Paris in observing the eclipses of Jupiter's moons. He converted them into so many signal-lamps, quenched when they plunged into the shadow of the planet, and relighted when they emerged from the shadow. They enabled him to prove that light requires time to pass through space, and to assign to it the astounding velocity of 190,000 miles a second. Then came the English astronomer, Bradley, who noticed that the fixed stars did not really appear to be fixed, but describe in the heavens every year a little orbit resembling the earth's orbit. The result perplexed him, but Bradley had a mind open to suggestion, and capable of seeing, in the smallest fact, a picture of the largest. He was one day upon the Thames in a boat, and noticed that, as long as his course remained unchanged, the vane upon his mast-head showed the wind to be blowing constantly in the same direction, but that the wind appeared to vary with every change in the direction of his boat. "Here," as Whewell says, "was the image of his case. The boat was the earth, moving in its orbit, and the wind was the light of a star."

We may ask in passing, what, without the faculty which formed the "image," would Bradley's wind and vane have been to him? A wind and vane, and nothing more. You will immediately understand the meaning of Bradley's discovery. Imagine yourself in a motionless railway-train with a shower of rain descending vertically downward. The moment the train begins to move, the rain-drops begin to slant, and the quicker the train the greater is the obliquity. In a precisely similar manner the rays from a star vertically overhead are caused to slant by the motion of the earth through space. Knowing the speed of the train, and the obliquity of the falling rain, the velocity of the drops may be calculated; and knowing the speed of the earth in her orbit, and the obliquity of the rays due to this cause, we can calculate just as easily the velocity of light. Bradley did this, and the "aberration of light," as his discovery is called, enabled him to assign to it a velocity almost identical with that deduced by Roemer from a totally different method of observation. Subsequently Fizeau, employing not planetary or stellar distances, but simply the breadth of the city of Paris, determined the velocity of light: while after him Foucault—a man of the rarest mechanical genius—solved the problem without quitting his private room.

Up to his demonstration of the composition of white light, Newton had been everywhere triumphant — triumphant in the heavens, triumphant on the earth, and his subsequent experimental work is for the most part of immortal value. But infallibility is not the gift of man, and, soon after his discovery of the nature of white light, Newton proved himself human. He supposed that refraction and dispersion went hand in hand, and that you could not abolish the one without at the same time abolishing the other. Here Dolland corrected him. But Newton committed a graver error than this. Science, as I sought to make clear to you in our second lecture, is only in part a thing of the senses. The roots of phenomena are embedded in a region beyond the reach of the senses, and less than the root of the matter will never satisfy the scientific mind.

We find, accordingly, in this career of optics, the greatest minds constantly yearning to pass from the phenomena to their causes—to explore them to their hidden roots. They thus entered the region of theory, and here Newton, though drawn from time to time towards the truth, was drawn still more strongly towards the error, and made it his substantial choice. His experiments are imperishable, but his theory has passed away. For a century it stood like a dam across the course of discovery; but, like all barriers that rest upon authority, and not upon truth, the pressure from behind increased, and eventually swept the barrier away. This, as you know, was done mainly through the labors of Thomas Young, and his illustrious French fellow-worker Fresnel.

In 1808, Malus, looking through Iceland spar at the sun reflected from the window of the Luxembourg Palace in Paris, discovered the polarization of light by reflection. In 1811 Arago discovered the splendid chromatic phenomena which we have had illustrated by plates of gypsum in polarized light; he also discovered the rotation of the plane of polarization by quartz-crystals. In 1813 Seebeck discovered the polarization of light by tourmaline. That same year Brewster discovered those magnificent bands of color that surround the axes of biaxal crystals. In 1814 Wollaston discovered the rings of Iceland spar. All these effects, which, without a theoretic clue, would leave the human mind in a hopeless jungle of phenomena without harmony or relation, were organically connected by the theory of undulation. The theory was applied and verified in all directions, Airy being especially conspicuous for the severity and conclusiveness of his proofs. The most remarkable verification fell to the lot of the late Sir William Hamilton, of Dublin, a profound mathematician, who, taking up the theory where Fresnel had left it, arrived at the conclusion that, at four special points at the surface of the ether-wave in double-refracting crystals, the ray was divided not into two parts, but into an infinite number of parts; forming at these points a continuous conical envelope instead of two images. No human eye had ever seen this envelope when Sir William Hamilton inferred its existence. Turning to his friend Dr. Lloyd, he asked him to test experimentally the truth of his theoretic conclusion. Lloyd, taking a crystal of arragonite, and following with the most scrupulous exactness the indications of theory, cutting the crystal where theory said it ought to be cut, observing it where theory said it ought to be observed, found the luminous envelope which had previously been a mere idea in the mind of the mathematician.

Nevertheless this great theory of undulation, like many another truth, which in the long-run has proved a blessing to humanity, had to establish, by hot conflict, its right to existence. Great names were arrayed against it. It had been enunciated by Hooke, it had been applied by Huyghens, it had been defended by Euler. But they made no impression. And, indeed, the theory in their hands was more an analogy than a demonstration. It first took the form of a demonstrated verity in the hands of Thomas Young. He brought the waves of light to bear upon each other, causing them to support each other, and to extinguish each other at will. From their mutual actions he determined their lengths, and applied his determinations in all directions. He showed that the standing difficulty of polarization might be embraced by the theory. After him came Fresnel, whose transcendent mathematical abilities enabled him to give the theory a generality unattained by Young. He grasped the theory in its entirety; followed the ether into its eddies and estuaries in the hearts of crystals of the most complicated structure, and into bodies subjected to strain and pressure. He showed that the facts discovered by Malus, Arago, Brewster, and Biot, were so many ganglia, so to speak, of his theoretic organism, deriving from it sustenance and explanation. With a mind too strong for the body with which it was associated, that body became a wreck long before it had become old, and Fresnel died, leaving, however, behind him a name immortal in the annals of science.

One word more I should like to say regarding Fresnel. There are things, ladies and gentlemen, better even than science. There are matters of the character as well as matters of the intellect, and it is always a pleasure to those who wish to think well of human nature, when high intellect and upright character are combined. They were, I believe, combined in this young Frenchman. In those hot conflicts of the undulatory theory, he stood forth as a man of integrity, claiming no more than his right, and ready to concede their rights to others. He at once recognized and acknowledged the merits of Thomas Young. Indeed, it was he, and his fellow-countryman Arago, who first startled England into the consciousness of the injustice done to Young in the *Edinburgh Review*. I should like to read you a brief extract from a letter written by Fresnel to Young in 1824, as it throws a pleasant light upon the character of the French philosopher. "For a long time," says Fresnel, "that sensibility, or that vanity, which people call love of glory, has been much blunted in me. I labor much less to catch the suffrages of the public than to obtain that inward approval which has always been the sweetest reward of my efforts. Without doubt, in moments of disgust and discouragement, I have often needed the spur of vanity to excite me to pursue my researches. But all the compliments I have received from Arago, De la Place, and Biot, never gave me so much pleasure as the discovery of a theoretic truth, or the confirmation of a calculation by experiment."

This, ladies and gentlemen, is the core of the whole matter as regards science. It must be cultivated for its own sake, for the pure love of truth, rather than for the applause or profit that it brings. And now my occupation in America is wellnigh gone. Still I will bespeak your tolerance for a few concluding remarks in reference to the men who have bequeathed to us the vast body of knowledge of which I have sought to give you some faint idea in these lectures. What was the motive that spurred them on? what the prize of their high calling for which they struggled so assiduously? What urged them to those battles and those victories over reticent Nature which have become the heritage of the human race? It is never to be forgotten that not one of those great investigators, from Aristotle down to Stokes and Kirchhoff, had any practical end in view, according to the ordinary definition of the word "practical." They did not propose to themselves money as an end, and knowledge as a means of obtaining it. For the most part, they nobly reversed this process, made knowledge their end, and such money as they possessed the means of obtaining it.

We may see to-day the issues of their work in a thousand practical forms, and this may be thought sufficient to justify it, if not ennoble their efforts. But they did not work for such issues; their reward was of a totally different kind. We love clothes, we love luxuries, we love fine equipages, we love money, and any man who can point to these as the result of his efforts in life justifies these efforts before all the world. In America and England more especially he is a "practical" man. But I would appeal confidently to this assembly whether such things exhaust the demands of human nature? The very presence here for six inclement nights of this audience, embodying so much of the mental force and refinement of this great city, is an answer to my question. I need not tell such an assembly that there are joys of the intellect as well as joys of the body, or that these pleasures of the spirit constituted the reward of our great investigators. Led on by the whisperings of natural truth, through pain and self-denial, they often pursued their work. With the ruling passion strong in death, some of them, when no longer able to hold a pen, dictated to their friends the results of their labors, and then rested from them forever.

Could we have seen these men at work without any knowledge of the consequences of their work, what should we have thought of them? To many of their contemporaries it would have appeared simply ridiculous to see men, whose names are now stars in the firmament of science, straining their attention to observe an effect of experiment almost too minute for detection. To the uninitiated, they might well appear as big children playing with not very amusing toys. It is so to this hour. Could you watch the true investigator—

your Henry or your Draper, for example—in his laboratory, unless animated by his spirit, you could hardly understand what keeps him there. Many of the objects which rivet his attention might appear to you utterly trivial; and, if you were to step forward and ask him what is the *use* of his work, the chances are that you would confound him. He might not be able to express the use of it in intelligible terms. He might not be able to assure you that it will put a dollar into the pocket of any human being living or to come. That scientific discovery *may* put not only dollars into the pockets of individuals, but millions into the exchequers of nations, the history of science amply proves; but the hope of its doing so never was and never can be the motive power of the investigator.

I know that I run some risk in speaking thus before practical men. I know what De Tocqueville says of you. "The man of the North," he says, "has not only experience, but knowledge. He, however, does not care for science as a pleasure, and only embraces it with avidity when it leads to useful applications." But what, I would ask, are the hopes of useful applications which have drawn you so many times to this place in spite of snow-drifts and biting cold? What, I may ask, is the origin of that kindness which drew me from my work in London to address you here, and which, if I permitted it, would send me home a millionnaire? Not because I had taught you to make a single cent by science, am I among you to-night, but because I tried to the best of my ability to present science to the world as an intellectual good. Surely no two terms were ever so distorted and misapplied with reference to man in his higher relations as these terms useful and practical. As if there were no nakedness of the mind to be clothed as well as nakedness of the body—no hunger and thirst of the intellect to satisfy. Let us expand the definitions of these terms until they embrace all the needs of man, his highest intellectual needs inclusive. It is specially on this ground of its administering to the higher needs of the intellect, it is mainly because I believe it to be wholesome as a source of knowledge, and as a means of discipline, that I urge the claims of science this evening upon your attention.

But, with reference to material needs and joys, surely pure science has also a word to say. People sometimes speak as if steam had not been studied before James Watt, or electricity before Wheatstone and Morse; whereas, in point of fact, Watt and Wheatstone and Morse, with all their practicality, were the mere outcome of antecedent forces, which acted without reference to practical ends. This also, I think, merits a moment's attention. You are delighted, and with good reason, with your electric telegraphs, proud of your steam-engines and your factories, and charmed with the productions of photography. You see daily, with just elation, the creation of new forms of industry—

new powers of adding to the wealth and comfort of society. Industrial England is heaving with forces tending to this end, and the pulse of industry beats still stronger in the United States. And yet, when analyzed, what are industrial America and industrial England? If you can tolerate freedom of speech on my part, I will answer this question by an illustration. Strip a strong arm, and regard the knotted muscles when the hand is clenched and the arm bent. Is this exhibition of energy the work of the muscle alone? By no means. The muscle is the channel of an influence, without which it would be as powerless as a lump of plastic dough. It is the delicate unseen nerve that unlocks the power of the muscle. And, without those filaments of genius which have been shot like nerves through the body of society by the original discoverer, industrial America and industrial England would, I fear, be very much in the condition of that plastic dough.

At the present time there is a cry in England for technical education, and it is the expression of a true national want; but there is no cry for original investigation. Still without this, as surely as the stream dwindles when the spring dries, so surely will "technical education" lose all force of growth, all power of reproduction. Our great investigators have given us sufficient work for a time; but, if their spirit die out, we shall find ourselves eventually in the condition of those Chinese mentioned by De Tocqueville, who, having forgotten the scientific origin of what they did, were at length compelled to copy without variation the inventions of an ancestry who, wiser than themselves, had drawn their inspiration direct from Nature.

To keep society as regards science in healthy play, three classes of workers are necessary: Firstly, the investigator of natural truth, whose vocation it is to pursue that truth, and extend the field of discovery for the truth's own sake, and without reference to practical ends. Secondly, the teacher of natural truth, whose vocation it is to give public diffusion to the knowledge already won by the discoverer. Thirdly, the applier of natural truth, whose vocation it is to make scientific knowledge available for the needs, comforts, and luxuries of life. These three classes ought to coexist and interact. Now, the popular notion of science, both in this country and in England, often relates, not to science strictly so called, but to the applications of science. Such applications, especially on this continent, are so astounding—they spread themselves so largely and umbrageously before the public eye—as to shut out from view those workers who are engaged in the quieter and profounder business of original investigation.

Take the electric telegraph as an example, which has been repeatedly forced upon my attention of late. I am not here to attenuate in the slightest degree the services of those who, in England and America, have given the telegraph a form so wonderfully fitted for public use. They earned a great reward, and assuredly they have received it. But I should be untrue to you and to myself if I failed to tell you that, however high in particular respects their claims and qualities may be, practical men did not discover the electric telegraph. The discovery of the electric telegraph implies the discovery of electricity itself, and the development of its laws and phenomena. Such discoveries are not made by practical men, and they never will be made by them, because their minds are beset by ideas which, though of the highest value from one point of view, are not those which stimulate the original discoverer.

The ancients discovered the electricity of amber; and Gilbert, in the year 1600, extended the force to other bodies. Then followed other inquirers, your own Franklin among the number. But this form of electricity, though tried, did not come into use for telegraphic purposes. Then appeared the great Italian, Volta, who discovered the source of electricity, which bears his name, and applied the most profound insight and the most delicate experimental skill to its development. Then arose the man who added to the powers of his intellect all the graces of the human heart, Michael Faraday, the discoverer of the great domain of magneto-electricity. Œrsted discovered the deflection of the magnetic needle, and Arago and Sturgeon the magnetization of iron by the electric current. The voltaic circuit finally found its theoretic Newton in Ohm, while Henry, of Princeton, who had the sagacity to recognize the merits of Ohm while they were still decried in his own country, was at this time in the van of experimental inquiry.

In the works of these men you have all the materials employed at this hour in all the forms of the electric telegraph. Nay, more; Gauss, the celebrated astronomer, and Weber, the celebrated natural philosopher, both professors in the University of Göttingen, wishing to establish a rapid mode of communication between the observatory and the physical cabinet of the university, did this by means of an electric telegraph. The force, in short, had been discovered, its laws investigated and made sure, the most complete mastery of its phenomena had been attained, nay, its applicability to telegraphic purposes demonstrated, by men whose sole reward for their labors was the noble joy of discovery, and before your practical men appeared at all upon the scene.

Are we to ignore all this? We do so at our peril. For I say again, that, behind all your practical applications, there is a region of intellectual action to which practical men have rarely contributed, but from which they draw all their supplies. Cut them off from this region, and they become eventually helpless. In no case is the adage truer, "Other men labored, but ye are entered into their labors," than in the case of the discoverer and the applier of natural truth. But now a word on the other side. While I say that practical men are not the men to make the necessary antecedent discoveries, the cases are rare in which the discoverer knows how to turn his labors to practical account. Different qualities of mind and different habits of thought are needed in the two cases; and, while I wish to give emphatic utterance to the claims of those whose claims, owing to the simple fact of their intellectual elevation, are often misunderstood, I am not here to exalt the one class of workers at the expense of the other. They are the necessary supplements of each other; but remember that one class is sure to be taken care of. All the material rewards of society are already within their reach; but it is at our peril that we neglect to provide opportunity for those studies and pursuits which have no such rewards, and from which, therefore, the rising genius of the country is incessantly tempted away.

Pasteur, one of the most eminent members of the Institute of France, in accounting for the disastrous overthrow of his country and the predominance of Germany in the late war, expresses himself thus: "Few persons comprehend the real origin of the marvels of industry and the wealth of nations. I need no further proof of this than the employment, more and more frequent in official language, and in writing of all sorts, of the erroneous expression *applied science*. The abandonment of scientific careers by men capable of pursuing them with distinction was recently complained of in the presence of a minister of the greatest talent. This statesman endeavored to show that we ought not to be surprised at this result, because *in our day the reign of theoretic science yielded place to that of applied science*. Nothing could be more erroneous than this opinion, nothing, I venture to say, more dangerous, even to practical life, than the consequences which might flow from these words. They have rested on my mind as a proof of the imperious necessity of reform in our superior education. There exists no category of the sciences to which the name of applied science could be given. *We have science, and the applications of science*, which are united together as the tree and its fruit.

And Cuvier, the great comparative anatomist, writes thus upon the same theme: "These grand practical innovations are the mere applications of truths of a higher order, not sought with a practical intent, but which were pursued for their own sake, and solely through an ardor for knowledge. Those who applied them could not have discovered them; those who discovered them had no inclination to pursue them to a practical end. Engaged in the high regions whither their thoughts had carried them, they hardly perceived these practical issues, though born of their own deeds. These rising workshops, these peopled colonies, those ships which furrow the seas — this abundance, this luxury, this tumult — all this comes from discoverers in science, and it all remains strange to them. At the point where science merges into practice, they abandon it; it concerns them no more."

When the Pilgrim Fathers landed at Plymouth Rock, and when Penn made his treaty with the Indians, the new-comers had to build their houses, to chasten the earth into cultivation, and to take care of their souls. In such a community, science, in its more abstract forms, was not to be thought of. And, at the present hour, when your hardy Western pioneers stand face to face with stubborn Nature, piercing the mountains and subduing the forest and the prairie, the pursuit of science, for its own sake, is not to be expected. The first need of man is food and shelter; but a vast portion of this continent is already raised far beyond this need. The gentlemen of New York, Brooklyn, Boston, Philadelphia, Baltimore, and Washington, have already built their houses, and very beautiful they are; they have also secured their dinners, to the excellence of which I can also bear testimony. They have, in fact, reached that precise condition of well-being and independence when a culture, as high as humanity has yet reached, may be justly demanded at their hands. They have reached that maturity, as possessors of wealth and leisure, when the investigator of natural truth, for the truth's own sake, ought to find among them promoters and protectors.

Among the many grave problems before them they have this to solve, whether a republic is able to foster the highest forms of genius. You are familiar with the writings of De Tocqueville, and must be aware of the intense sympathy which he felt for your institutions; and this sympathy is all the more valuable, from the philosophic candor with which he points out, not only your merits, but your defects and dangers. Now, if I come here to speak of science in America in a critical and captious spirit, an invisible radiation from my words and manner will enable you to find me out, and will guide your treatment of me to-night. But, if I, in no unfriendly spirit—in a spirit, indeed, the reverse of unfriendly—venture to repeat before you what this great historian and analyst of democratic institutions said of America, I am persuaded that you will hear me out. He wrote some three-and-twenty years ago, and perhaps would not write the same to-day; but it will do nobody any harm to have his words repeated, and, if necessary, laid to heart. In a work published in 1850, he says: "It must be confessed that, among the civilized peoples of our age, there are few in which the highest sciences have made so little progress as in the United States."[1] He declares his conviction that, had you been

---

[1] Il faut reconnaître, que parmis les peuples civilisés des nos jours, il en est peu chez qui les hautes sciences aient fait moins de

alone in the universe, you would speedily have discovered that you cannot long make progress in practical science, without cultivating theoretic science at the same time. But, according to De Tocqueville, you are not thus alone. He refuses to separate America from its ancestral home; and it is here, he contends, that you collect the treasures of the intellect, without taking the trouble to create them.

De Tocqueville evidently doubts the capacity of a democracy to foster genius as it was fostered in the ancient aristocracies. "The future," he says, "will prove whether the passion for profound knowledge, so rare and so fruitful, can be born and developed so readily in democratic societies as in aristocracies. As for me," he continues, "I can hardly believe it." He speaks of the unquiet feverishness of democratic communities, not in times of great excitement, for such times may give an extraordinary impetus to ideas, but in times of peace. There is then, he says, "a small and uncomfortable agitation, a sort of incessant attrition of man against man, which troubles and distracts the mind without imparting to it either animation or elevation." It rests with you to prove whether these things are necessarily so—whether the highest scientific genius cannot find in the midst of you a tranquil home.

progrès qu'aux Etats-Unis, on qui aient fourni moins de grands artists, de poètes illustres, et de célèbres écrivians. (De la Démocratie en Amérique, etc., tome ii., p. 36.)

I should be loath to gainsay so keen an observer and so profound a political writer, but, since my arrival in this country, I have been unable to see anything in the constitution of society to prevent a student with the root of the matter in him from bestowing the most steadfast devotion on pure science. If great scientific results are not achieved in America, it is not to the small agitations of society that I should be disposed to ascribe the defect, but to the fact that the men among you who possess the endowments necessary for scientific inquiry are laden with duties of administration or tuition so heavy as to be utterly incompatible with the continuous and tranquil meditation which original investigation demands. It may well be asked whether Henry would have been transformed into an administrator, or whether Draper would have forsaken science to write history, if the original investigator had been honored as he ought to be in this land? I hardly think they would. Still I do not think this state of things likely to last. In America there is a willingness on the part of individuals to devote their fortunes, in the matter of education, to the service of the commonwealth, which is without a parallel elsewhere; and this willingness requires but wise direction to enable you effectually to wipe away the reproach of De Tocqueville.

Your most difficult problem will be not to build institutions, but to make men; not to form the body, but to find the spiritual embers which shall kindle within that body a living soul. You have scientific genius among you; not sown broadcast, believe me, but still scattered here and there. Take all unnecessary impediments out of its way. Drawn by your kindness I have come here to give these lectures, and, now that my visit to America has become almost a thing of the past, I look back upon it as a memory without a stain. No lecturer was ever rewarded as I have been. From this vantage-ground, however, let me remind you that the work of the lecturer is not the highest work; that in science the lecturer is usually the distributor of intellectual wealth amassed by better men. It is not solely, or even chiefly, as lecturers, but as investigators, that your men of genius ought to be employed. Keep your sympathetic eye upon the originator of knowledge. Give him the freedom necessary for his researches, not overloading him either with the duties of tuition or of administration, not demanding from him so-called practical results—above all things, avoiding that question which ignorance so often addresses to genius, "What is the use of your work?" Let him make truth his object, however unpractical for the time being that truth may appear. If you cast your bread thus upon the waters, then be assured it will return to you, though it may be after many days.

APPENDIX.[1]

At a banquet given in honor of Prof. TYNDALL, at Delmonico's, New York, February 4, 1873, the Chairman, Hon. Wm. M. Evarts, gave the toast of the evening: "The health of Prof. TYNDALL, our honored guest."

Prof. TYNDALL, rising to respond, said:

I happened to know, sir, that my eminent friends, Profs. Henry and Agassiz, had been asked to accept the position which you are so good as to honor on the present occasion, and I knew you would not misinterpret me if I prefaced what I had to say by an expression of regret that overpowering reasons prevent either of them from being present here to-night.

But I fear that in honesty I must now go further than this, and at the risk of apparent rudeness avow the wish that one of the gentlemen alluded to, and not Mr. Evarts, were presiding here to-night. For neither of them would, I apprehend, have given me the task now before me, of following with my commonplace utterances a speech so spar-

[1] From "The Proceedings of the Tyndall Banquet." D. Appleton & Co.

kling with wit and humor as that which we have been privileged to hear, and which required for its production a discipline in public speaking hardly to be assumed in the case of my other two friends.

I have also, sir, to regret the withdrawal of an object from our vicinity, from which I might have derived a momentary inspiration. In front of you, sir, a moment ago, towered a noble Alpine peak, with men upon its ledges, suspending others by ropes adown its crags. Thus before me stood all the pomp and circumstance of mountain-climbing, and the thoughts of liberty aroused by such an object might have helped to give me freedom of speech; but alas! the waiter, not knowing that he was virtually stealing the thread of my discourse, has removed this source of inspiration.

There is, however, one point in your speech, sir, which requires simple honesty and little wit to respond to. That point is symbolized by those festooned flags of America and England which I now see before me. You spoke, sir, of the sympathy existing between the intellect of the United States and that of England, and of the smallness of our differences compared with the vast area in which we coincide. Coming from you, sir, these words had a peculiar weight and worth to me. I am persuaded that they are not the words of mere conventional compliment, but that they embody your convictions. And I am equally persuaded that they are the expression of a fact which will become more and more prominent as time rolls on, and as international knowledge is increased.

During my four months' residence in the United States I have not heard a single whisper hostile to England; and this accounts for a certain change of feeling, accompanied by a certain change of expression on my part. Among the motives which prompted me to come here was this: I thought it possible that a man withdrawn from the arena of politics, and who had been fortunate enough to gain a measure of the good-will of the American people, might do something to soften the asperities arising out of political differences. I said something bearing upon this point in Boston; but my references to it have grown more and more scanty, until in the three cities last visited they disappeared altogether. And this was not because I had the subject less at heart, but because I saw that reference to it was unnecessary and out of place, resembling, as Mr. Emerson would say, the sound of a scythe in December when there is nothing to mow. We are not angels on either side of the Atlantic, nor am I aware that we desire to be angels; but as men I believe there exists a strength of brotherhood between us which, when liberated from the mists of ignorance, will weld the two nations almost as closely together as the various parts of your own vast community are welded. "It behooves us all to forward this result." These were the last words of my excellent friend Mr. Russell Gurney, as I shook his hand on the railway platform at Washington.

And now to my science. I should like to have seen Prof. Henry here to-night, because, before I was invested with my scientific swaddling-clothes he was in the van of experimental inquiry. From a scientific point of view it is to be deplored that such a man has been transformed from an investigator into an administrator. From the same point of view it is to be regretted, if he will allow me to connect his name with such an expression of regret, that an eminent friend of mine now at this table, Dr. Draper, however high his gifts in his new vocation may be, and assuredly they are very high, has been deflected from scientific research to the writing of history. With regard to Prof. Agassiz, I had the pleasure of making his acquaintance in Geneva in 1859, and I am intimately acquainted with the scenes of his scientific action in the Alps. I had also the privilege of visiting him at Cambridge, of going with him through the noble institution of which he is the living head, and of witnessing his mode of working with his pupils. And I can only say, that if from these pupils men do not spring, able and willing to carry forward the sacred fire of original investigation, it will not be the fault of the master.

The interest shown in the labors to which you, sir, have so kindly referred is not, in my opinion, the creation of the hour. Every such display of public sympathy must have its prelude, during which men's minds are prepared, a desire for knowledge created, an intelligent curiosity aroused. Then in the nick of time comes a person, who, though but an accident, touches a spring which permits tendency to flow into fact, and public feeling to pass from the potential to the actual. The interest displayed has really been the work of years, and the chief merit rests with those who were wise enough to discern that, as regards physics, the detent might be removed and the public sympathy for that department of science permitted to show itself. Among the foremost of those who saw this must be reckoned my indefatigable friend Prof. Youmans. In no other way can I account for my four months' experience in the United States. The soil had been prepared and the good seed sown long before I came among you. And it is on this belief that the subject has a root deeper than the curiosity of the hour that I found my hopes of its not passing rapidly from the public mind.

It would be a great thing, sir, for this land of incalculable destinies to supplement its achievements in the industrial arts by those higher investigations from which our mastery over Nature and over industrial art has been derived, and which, when applied in a true catholic spirit to man himself, will assuredly render him healthier, stronger, purer, nobler than he now is. To no other country is the cultivation of science in its highest forms of more importance than to yours. In no other country would it exert a more benign and elevating influence. What, then, is to be done toward so desirable a consummation? Here I think you must take counsel of your leading scientific men; and they are not unlikely to recommend something of this kind. I think, as regards physical science, they are likely to assure you that it is not what I may call the statical element of *buildings* that you require so much as the dynamical element of *brains*. Making use as far as possible of existing institutions, let chairs be founded, sufficiently but not luxuriously endowed, which shall have original research for their main object and ambition. With such vital centres among you, all your establishments of education would feel their influence; without such centres, even your primary instruction will never flourish as it ought. I would not, as a general rule, wholly sever tuition from investigation, but, as in the institution to which I belong, the one ought to be made subservient to the other. The Royal Institution gives lectures—indeed, it lives in part by lectures, though mainly by the contributions of its members, and the bequests of its friends. But the main feature of its existence—a feature never lost sight of by its wise and honorable Board of Managers—is that it is a school of research and discovery. Though a by-law gives them the power to do so, for the twenty years during which I have been there no manager or member of the Institution has ever interfered with my researches. It is this wise freedom, accompanied by a never-failing sympathy, extended to the great men who preceded me, that has given to the Royal Institution its imperishable renown.

I have said that I could not wholly sever tuition from investigation, and I should like to add one word to this remark. In your chairs of investigation let such work as that in which I have been lately engaged be reduced to a minimum. Look jealously upon the investigator who is fond of wandering from his true vocation to appear on public platforms. The practice is absolutely destructive of original work of a high order. Now and then the discoverer, when he has any thing important to tell, may appear with benefit to himself and the world. But as a general rule he must leave the work of public lecturing to others. This may appear to you a poor return for the plaudits with which my own efforts have been received; but these efforts had a special aim. My first duty toward you, moreover, is to be true, and what I say here is the inexorable truth.

As to the source of the funds necessary for founding the chairs to which I have referred, it is not for me to offer an opinion. Without raising the disputed question of State aid, in this country it is possible to do a great deal without it. As I said in my lectures, the willingness of American citizens to throw their fortunes into the cause of public education is without a parallel in my experience. Hitherto their efforts have been directed to the practical side of science, and this is why I sought in my lectures to show the dependence of practice upon principles. On the ground, then, of mere practical, material utility, pure science ought to be cultivated. But assuredly among your men of wealth there are those willing to listen to an appeal on higher grounds, to whom, as American citizens, it will be a pride to fashion American men so as to enable them to take their places among those great ones mentioned in my lectures. Into this plea I would pour all my strength. Not as a servant of Mammon do I ask you to take science to your hearts, but as the strengthener and enlightener of the mind of man.

Might I now address a word or two to those who in the ardor of youth feel themselves drawn toward science as a vocation. They must, if possible, increase their fidelity to original research, prizing far more than the possession of wealth an honorable standing in science. They must, I think, be prepared at times to suffer a little for the sake of scientific righteousness, not refusing, should occasion demand it, to live low and lie hard to achieve the object of their lives. I do not here urge any thing upon others that I should have been unwilling to do myself when young. Let me give you a line of personal history. In 1848, wishing to improve myself in science, I went to the University of Marburg—the same old town in which my great namesake, when even poorer than myself, published his translation of the Bible. I lodged in the plainest manner, in a street which, perhaps, bore an appropriate name while I dwelt upon it. It was called the *Ketzerbach*—the heretic's brook—from a little historic rivulet running through it. I wished to keep myself clean and hardy in an economical way, so I purchased a cask and had it cut in two by a carpenter. Half that cask, filled with spring water overnight, was placed in my small bedroom, and never during the years that I spent there, in winter or in summer, did the clock of the beautiful Elizabethkirche, which was close at hand, finish striking the hour of six in the morning before I was in my tub. For a good portion of the time I rose an hour and a half earlier than this, working by lamp-light at the differential calculus when the world was slumbering round me. And I risked this breach in my pursuits, and this expenditure of time and money, not because I had any definite prospect of material profit in view, but because I thought the cultivation of the intellect important—because, moreover, I loved my work, and entertained the sure and certain hope that, armed with knowledge, one can successfully fight one's way through the world.

It is with the view of giving others the chance that I then enjoyed that I propose to devote the surplus of the money which you have so generously poured in upon me, to the education of young philosophers in Germany.[1] I ought

---

[1] Prof. Tyndall's receipts from the lectures in the several cities were as follows:

Boston, six lectures	$1,500
Philadelphia, six lectures	3,000
Baltimore, three lectures	1,000
Washington, six lectures	2,000
New York, six lectures	8,500
Brooklyn, six lectures	6,100
New Haven, two lectures	1,000
Total	$23,100

Of this amount, the surplus above expenses, amounting to upward of $13,000, was conveyed, by an article of trust, to the charge of a committee, consisting of Prof. Joseph Henry, Gen. Hector Tyndale, and Prof. E. L. Youmans, who are authorized to expend the interest in aid of students who devote themselves to original researches.—PUBLISHERS' NOTE.

ADDRESS AT THE FAREWELL BANQUET. 191

not, for their sake, to omit one additional motive by which I was upheld at the time here referred to—that was, a sense of duty. Every young man of high aims must, I think, have a spice of this principle within him. There are sure to be hours in his life when his outlook will be dark, his work difficult, and his intellectual future uncertain. Over such periods, when the stimulus of success is absent, he must be carried by his sense of duty. It may not be so quick an incentive as glory, but it is a nobler one, and gives a tone to character which glory cannot impart. That unflinching devotion to work, without which no real eminence in science is now attainable, implies the writing at certain times of the stern resolve upon the student's character: "I work not because I like to work, but because I ought to work." In science, however, love and duty are sure to be rendered identical in the end.

And now I have reached the point where I am forced to qualify the expression of the pleasure which this visit has given me. With regard to its positive side—to work actually done, and to the reception of that work—nothing can be added to my cup of satisfaction. My only drawback relates to work undone; for I carry home with me the consciousness of having been unable to respond to the invitations of the great cities of the West, thus, I fear, causing in many cases disappointment. But the character of my lectures, the weight of instrumental appliances which they involved, and the fact that every lecture required two days' possession of the hall—a day of preparation and a day of delivery—entailed heavy loss of time, and often very severe labor. The mere need of rest would be sufficient to cause me to pause here; but added to this is the fact that every mail from England brings me intelligence of works suspended and duties postponed through my absence. We have an honorary secretary who has devoted the best years of an active professional life and the best energies of a strong man, to the interests of the Royal Institution. And I would say in passing that, if ever you found any thing similar to our institution in the United States, the heartiest wish that I could offer for its success would be, that it may be served and aided with the same self-sacrificing love and fidelity which have characterized the service rendered to the Royal Institution; and by none more devotedly than by its present eminent honorary secretary, Dr. Bence Jones. But he, on whom I might rely, and who would willingly have taken my place, is now smitten down by a distressing illness; and, though other friends are willing to aid me in all possible ways, there can be no doubt as to my line of duty. I ought to be at home. I ask my friends in the West to take these things into consideration; I ask them to believe that, if it lay within the limits of my power, I should be among them; I ask of them to think of me, not with bitterness or disappointment, not as one insensible to their kindness, but as a friend who, with a warmth commensurate with their own, would comply with all their wishes if he could.

One other related point deserves mention. On quitting England I had no intention of publishing the lectures I have given here, and, except a fragment or two, they were wholly unwritten when I arrived in this city. Since that time, besides lecturing in New York, Brooklyn, and New Haven, the lectures have been written out. No doubt many evidences of the rapidity of their production will appear; but I thought it due to those who listened to them with such unwavering attention, as also to those who wished to hear them, but were unable to do so, to leave them behind me in an authentic form. The constant application which this work rendered necessary has cut me off from many social pleasures; it has also prevented me from making myself acquainted with the working of institutions in which I feel a deep interest; it has prevented me from availing myself of the generous hospitality offered to me by your clubs. In short, it has made me an unsociable man. But, finding social pleasure and hard work incompatible, I took the line of devoting such energy as I could command, not to the society of my intimate friends alone, but to the people of the United States.

And now, gentlemen, all is nearly over, and in a day or two I quit these shores. I read a day or two ago an article in the *Galaxy*, in which the writer, who had been in England, and who had had what you call "a good time" in England, spoke nevertheless of the deep pleasure of reaching his own home. The words struck a sympathetic chord within me. And it is a curious psychological fact, that this home-yearning, in my case, is not only unopposed, but is actually aided by the feeling that since I came to this country America has been a home to me. It is not a case of two opposing attractions, but a case in which, one of the attractions being satisfied, I am left not only free to be acted on, but more ready to be acted on by the other. Were there any lingering doubt as to my visit at the bottom of my mind; did I feel that I had blundered—and with the best and purest intentions I might, through an error of judgment, have blundered—so as to cause you discontent, I should now be wishing to abolish the doubt or to repair the blunder. This would be so much withdrawn from the pleasurable thought of home. But there is no drawback of this kind; and, therefore, as I have said, the fulness of my content here enhances the prospective pleasure of meeting my older friends. By some means or other, gentlemen, the people of this country have begotten and fostered a strange confidence in me toward them. I feel as if I, a simple scientific student, who never taught the world to be a cent richer, who merely sought to present science to the world as an intellectual good, am leaving, not a group of friends merely, not merely a friendly city, but a friendly continent here behind me. The very disappointment of the West I take as a measure of the West's friendship. Tested and true hearts are awaiting me at the other side, and, thinking of them and, you the pure cold intellect is for the moment deposed, and what is called the "human heart" becomes master of the situation; but lest it, in the waywardness of strong emotion, should utter any thing which the reënthroned intellect of to-morrow might condemn, I will pause here—hoping, not for the entire consummation, for that would be a hope too daring, but hoping, as the generations pass, that the attachment which binds me to America, on the one side, and "the Old Country," on the other, may be more and more approached and realized by the nations themselves.

THE END.